THE MAGIC GARDEN OF
GEORGE B
AND OTHER LOGIC PUZZLES

The Magic Garden of

GEORGE B

and Other Logic Puzzles

Raymond Smullyan

Indiana University, USA

World Scientific

NEW JERSEY · LONDON · SINGAPORE · BEIJING · SHANGHAI · HONG KONG · TAIPEI · CHENNAI

Published by

World Scientific Publishing Co. Pte. Ltd.

5 Toh Tuck Link, Singapore 596224

USA office: 27 Warren Street, Suite 401-402, Hackensack, NJ 07601

UK office: 57 Shelton Street, Covent Garden, London WC2H 9HE

Library of Congress Cataloging-in-Publication Data
Smullyan, Raymond M.
   The magic garden of George B and other logic puzzles / by Raymond Smullyan (Indiana University, USA).
      pages cm
   Includes bibliographical references.
   ISBN 978-981-4675-05-5 (hardcover : alk. paper) -- ISBN 978-981-4678-55-1 (pbk : alk paper)
   1. Logic puzzles. 2. Mathematical recreations. I. Title. II.Title: Logic puzzles.
   QA95.S49965 2015
   793.74--dc23
                                    2015002792

British Library Cataloguing-in-Publication Data
A catalogue record for this book is available from the British Library.

Cover image: The picture of the man is George Boole (1815–1864)

Typeset by Stallion Press
Email: enquiries@stallionpress.com

Printed in Singapore

Foreword

The author of this highly entertaining book is well known, not only as a mathematician and magician, but particularly for his popular puzzle books, which introduce deep mathematical ideas to the general reader. The remarkable thing about these books is that they are of interest both to those who are already familiar with formal logic and to those who are not!

This book is no exception, although it is written primarily for those who do *not* have such a background – everything here is developed from scratch. The book's main purpose is to introduce the general reader to the fascinating subject of *Boolean logic*, known also as *Boolean algebra*, though it is totally unlike the algebra one learns in high-school, since it deals not with numbers, but rather with *logical* concepts such as truth and falsity. The subject is vital nowadays for the whole of computer science, as well as for other fields such as electrical engineering and artificial intelligence. You will find such weird looking equations as 1+1=0, but here the symbols "1" and "0" do not stand for numbers, but for other things, depending on the applications – for example, in logic they respectively stand for *truth* and *falsity*, whereas for electrical engineering "1" stands for *on* and "0" for *off*. As for "+", you will see what it stands for.

The subject has an interesting historical background: Among the ancients, Euclid stands out as introducing logical reasoning into Geometry, and Aristotle codified logic in a more formal manner. But one of the main pioneers of *modern* logic – that is, *symbolic* logic – was surely George Boole (1815-1865), who in his famous book *Laws of Thought* put logic itself into an algebraic form. In this sense, the book is an inspiring introduction to contemporary logic, the subject that Boole helped to forge.

The present delightful book, in the hands of a magician, amuses everyone with the deep ideas behind Boole's thought, made refreshingly simple as it promenades through some fantastic places called *Boolean Gardens* and *Boolean Islands*, whose inhabitants lie on some days and tell the truth on others, but always in strict accordance with Boole's laws.

Smullyan starts out very gently with recreational logic puzzles, and ends up by showing how to unify theories of these Boolean gardens and islands, propositional logic and the Boolean theory of sets into a general abstract theory – and the magic happens in the fact that the reader does not even perceive being the one who is performing such a grandiose abstraction!

Walter Carnielli
Editor, *Contemporary Logic*
Campinas, June 2006

Preface

Here is a remarkable problem: Imagine a garden of magic flowers that can change color from day to day. On any one day, a flower is either blue the entire day or red the entire day, but can change from one day to another. Given any flower A and any flower B, there is a flower C that is red on all and only those days on which A and B are both blue. Also, we are given that for any two *distinct* flowers A and B, there is at least one day on which A and B are of different colors. Now suppose that the number of flowers is somewhere between 200 and 500. How many flowers are in the garden?

Amazingly enough, the problem actually has a unique solution! Doesn't this surprises you? The solution, which reads like a detective story, unravels a host of facts which lead to the very heart of the subject known as *Boolean logic*, with such weird equations as $1 + 1 = 0$. This subject is vital these days for the entire field of computer science and artificial intelligence. Book II of this volume provides a guided tour of this fascinating subject for the general reader with no necessary background in logic or algebra. It can be read quite independently of Book I, which consists of miscellaneous logic and arithmetic puzzles for the general puzzle lovers. But, as I have already indicated, none of this is necessary for Book II. Indeed, Book I and Book II could have been published as separate volumes, but I figured that I would probably have a wider audience if the two were combined into one volume and thus be of interest both to puzzle fans and to the general reader who is curious to find out what the intriguing subject of Boolean logic is all about.

Elka Park, New York
September, 2005

Contents

BOOK I
It's All a Question of Logic!

Chapter 1

Puzzles or Monkey Tricks?

Before turning to the logic of lying and truth telling, I would like to entertain you with some miscellaneous items–puzzles, jokes, swindles, etc. Anything goes! This chapter is a free-for-all! (solutions to puzzles are given at the end of chapters).

A – Two Logical Hustles

1 – A Logical Three Card Monte — Many of you have seen street vendors playing a game known as *Three Card Monte*, in which the operator shows an ace and two other cards, then puts them face down on the table, mixes them up, and you are to bet on which one is the ace. Well, here is what might be called a *logical* version of the game.

There are 3 cards face down on the table; each one is either red or black (no joker). At most one of them is red. Also, one and only one of them is an ace, but whether a red ace or a black ace is not given. On the back of each card is written a sentence, and if the card is red, the sentence is true, but if the card is black, the sentence is false. Here are what the backs say:

Card 1	Card 2	Card 3
This card is not the ace.	This card is the ace.	Card 1 is not the ace.

Which of the three cards is the ace, and is the ace red or black?

2 – Find the Joker — Here is another "logical" three card monte: Three cards are face down on the table; one is red, one is black, and one is the joker. Again a sentence is written on the back of each card. The red card has a true sentence, the black card has a false sentence, and the sentence on the joker could be either true or false. Here are the cards:

Card 1	Card 2	Card 3
Card 3 is the joker	Card 1 is black	This card is the joker

Which card is the Joker? Also, which card is red and which is black?

B – What About These?

3 – An Old Timer — A man held two American coins in his hand which added up to 30 cents, yet one of them was not a nickel. What coins were they?

4 – What Is the Explanation? — In 1920, a man went into a bar and needed a nickel for a telephone call. He asked for change of a dollar. The bartender said: "I'm sorry, I can't give you change of a dollar. Do you by any chance have a five dollar bill?"

"Why yes", said the man.

"I can change that", said the bartender, which he did, and the man could then make his telephone call.

What is the explanation? [This is a genuine puzzle, not a monkey trick!]

5 – A man once threw a golf ball which went a short distance, came to a dead stop, reversed its motion and then went the opposite way. He didn't bounce it, nor did he hit it, or tie anything to it. What is the explanation?

6 – A man was driving along a thoroughfare. The headlights of his car were broken, and by a curious coincidence, there was a power shortage in the city and none of the street lamps could operate. Also, there was no moon out. A few hundred yards in front of him was a pedestrian crossing the street. Somehow, the driver was aware of the pedestrian and braked to a halt. How did he know that the pedestrian was there?

7 – Here is a puzzle (which some of you may well know) that I fell for! A boat has a metal ladder coming down the side. It has 6 rungs spaced 1 foot apart. At low tide, the water came up to the second rung from the bottom. Then the water rose two feet. Which rung did it then hit?

8 – How Is Your Arithmetic? — In a certain town, 13 percent of the inhabitants have unlisted phone numbers, and none of them have more than one phone number. Well, one day a statistician visited the town and picked 1,300 names at random from the phone book. Roughly, how many of them would you expect to have unlisted phone numbers?

9 – Could They Both Be Right? — I once visited two brothers named *Arthur* and *Robert*. Arthur told me that he had twice as many girl friends as Robert.

Then, to my surprise, Robert told me that he had twice as many girl friends as Arthur. Could they both be right?

10 – What is the easiest method to tell if a bird is male or female?

C – What Are the Odds?

11 – Suppose you write down a number from 1 to 20, and I write down a number from 1 to 20. What is the probability that your number will be higher than mine?

12 – What is the probability that there are at least two Americans who have exactly the same number of American friends? [I am assuming that friendship is mutual–if John is a friend of Bill, then Bill is also a friend of John, and I am not counting a person as his or her own friend. Also, I am *not* assuming that every American has at least one friend, as that would make it too easy! I *am* assuming that there are at least two Americans. The rest is pure logic!]

13 – Suppose I bet someone that he or she cannot tell me which president's picture is on a ten dollar bill without looking at the bill. Now, without *your* looking at the bill, can you tell me the probability that I will win the bet?

14 – A statistician once visited a convention of physicists and chemists, 50 scientists all told. He observed that whichever two of them were picked at random, one of them was bound to be a chemist. Now, suppose that just one of them is picked at random. What is the probability that he is a chemist?

15 – Three statisticians once visited a garden of red, blue, yellow, and white flowers. One of the statisticians observed that whichever four flowers were picked, one of them was bound to be red. Another observed that whichever four were picked, at least one of them was bound to be blue. The third statistician observed that whichever four were picked at least one was bound to be yellow. Now, if four are picked, what is the probability that at least one will be white?

* * *

Speaking of statisticians, there is the story that a statistician once told a friend that he never travels by air, because he computed the probability that there be a bomb on the plane, and although the probability was low, it was too high for his comfort. Two weeks later, the friend met the statistician on a plane and asked him why he had changed his theory. The statistician replied: "I didn't change my theory. It's just that I subsequently computed the probability that there be two bombs on the plane, and this probability is low enough for my comfort. So now I simply carry my own bomb".

D – Three Logic Puzzles

16 – A Foursome — Two married couples were having tea together. Two of the four were French and the other two were German. One of the four was a chemist, one a doctor, one a lawyer, and one a writer. The doctor is a Frenchman. The writer is a French woman, and her husband is a lawyer. Mr. Schmidt is German. What is the profession of Mrs. Schmidt?

17 – Three Sisters — Of three sisters, two are married, two are blond and two are secretaries. The one who is not blond is not a secretary, and the one who is not a secretary is single. How many, if any, are blond, married secretaries?

18 – Three Other Sisters — There is another set of three sisters named Arlene, Beatrice, and Cynthia. Arlene is unmarried. Beatrice is shorter than the youngest of the three. The oldest of the three is married and is also the tallest. Which of the three is the oldest and which is the youngest?

E – Theme and Variations

19 – A Probability Dilemma — Now comes a *very* serious problem!

A bag contains just one marble, which is either black or white, and with equal probability. A second black marble is thrown into the bag. The bag is then shaken, and a marble is withdrawn which turns out to be black. What is the probability that the remaining marble is black?

I shall now give two arguments which lead to completely incompatible conclusions, and the problem is to figure out what is wrong!

Argument 1 — Before knowing that the removed marble was black, there are the following four equally likely possible cases:

Case 1 - The marble originally in the bag was white and it was the marble that was removed.

Case 2 - The original marble was white and the added black marble was the one that was removed.

Case 3 - The original marble was black and it was the marble that was removed.

Case 4 - The original marble was black and the added black marble was the one that was removed.

Now, once we know that the removed marble was black, then Case 1 can no longer hold, and so one of the last three cases must hold and each is equally probable. In two of the cases (Cases 3 and 4) a black marble remains, and only in Case 2 does a white marble remain. Therefore, the probability that the remaining marble is black is two-thirds.

Argument 2 — It is obvious that if the original marble is white, then the remaining marble is white (since it wasn't removed) and if the original marble is black, then the remaining marble must be black (either the original one or the added one). Thus, the color of the marble in the bag is not changed by adding a black marble and then removing a black marble. Therefore, since the probability is one-half that the original marble was black and the probability hasn't changed by adding and removing a black marble, the probability that the remaining marble is black is also one-half, not two-thirds!

Well, something is clearly wrong with one of the arguments. Which one, and where is the fallacy?

20 – **A Variant** — Suppose that in the last problem, instead of being told that a marble was removed from the bag at random, we are told that someone looked into the bag and deliberately removed a black marble. Would that change the answer?

21 – **Another Variant** — Suppose, as in Problem 16, a marble was removed at random, but we are now not told whether it was black or white. Then what is the probability that the remaining marble is black?

F – A Special Problem

22 – **Who Is What?** — David and Edward are brothers. One is a programmer and the other is an engineer. David is exactly 26 weeks older than Edward who was born in August. The programmer, who was born in January, was 54 years old in 1998. Which of the two brothers is the engineer? [This is a genuine puzzle, not a monkey trick.]

* * *

Speaking of programmers and engineers, there is the story going around about an engineer and a programmer who were sitting side by side on an airplane. The programmer asked the engineer whether he would like to play a game. "No, I want to sleep", said the engineer.

"It's a very good game"! said the programmer.

"No, I want to sleep"!

The programmer then said: "You ask me a question and if I don't know the answer, I pay you five dollars. Then I ask you a question, and if you don't know the answer, then *you* pay *me* five dollars".

"Nah, I want to sleep".

"I'll tell you what: If you don't know the answer, you pay me five dollars, but if I don't know the answer, I'll pay you *fifty* dollars"!

"Oh, all right".

The engineer went first: "What goes up the hill with three legs and comes down with four?" The programmer took out his lap top computer and worked on the reference for about an hour. He then shook his head and handed the engineer fifty dollars. The engineer said nothing and put the fifty dollars in his pocket. The programmer, a bit miffed, said: "Well, what's the answer?" At which the engineer handed him five dollars.

SOLUTIONS

1 – Since, at most, one of the three cards is red, then at most one of the three sentences is true. Now, the sentence on Card 1 and Card 3 agree, hence they are either both true or both false. Since they cannot both be true, they are both false. Since the sentence in Card 1 is false, then Card 1 is the ace–and it is black.

2 – Card 3 is obviously not the red one, since the sentence on the red card is true. If Card 2 is the red one, then Card 1 will be black (as Card 2 says), hence Card 3 would be the joker, hence Card 1 would have a true sentence, which a black card cannot have. Therefore, Card 2 cannot be red. This leaves Card 1 as the red one, and hence Card 3 is the joker. And so Card 1 is red, Card 2 is black and Card 3 is the joker.

3 – He had a nickel and a quarter. One of them (namely, the quarter) was not a nickel. [This is what I call a *monkey trick*!]

4 – The bartender had a 2 1/2 dollar gold piece (which was fairly prevalent in those days) and gave it to the customer, together with 2 dollar bills and a quarter, two dimes, and a nickel.

5 – He threw it straight up in the air.

6 – It was daytime.

7 – When I heard this, I said: "The obvious answer is the fourth rung from the bottom, but it is too obvious to be correct, but I can't see what's wrong with my arithmetic"! Well, the correct answer is the *second* rung, since the boat rises with the water.

8 – Here is another one I stupidly fell for! I replied "169" (which is 13 percent of 1300), but the correct answer is *zero*, since the names were taken from the telephone book!

9 – They were both right: Neither brother had any girl friends at the time, and twice zero is zero.

10 – You offer the bird some seed. If he eats it, then the bird is male. If she eats it, then it's a female.

"But", you might reply, "how do you know whether it's a he or a she"?

Answer: If it's male, then it's a he. If it's female, then it's a she. [I suspect this reasoning is slightly circular].

11 – The chances that both numbers are the same is one out of twenty, hence the chances that they are different is nineteen out of twenty. If they are different, the chances are even that your number is greater then mine, and so the total probability that your number is greater than mine is one half of nineteen out of twenty, which is 19 out of 40.

12 – The probability is 100%. Indeed, given *any* group of at least two people there must be at least two members who have exactly the same number of friends in the group, for suppose, say, that the group has 1,000 members. Now, there are exactly one thousand whole numbers less than 1,000–namely the numbers from zero to 999, and so the only way that no two members of the group can have a different number of friends (in the group) is that for each number from zero to 999 (inclusive), there is one member who has just that number of friends. Thus, one member must have zero friends, one must have 1 friend, one must have 2 friends, and so on, up to one having 999 friends. But this is not possible since if one member has 999 friends, then he has all the others as friends, hence all the others must have him as a friend, so none of them can have zero friends in the group. Therefore, at least two members must have exactly the same number of friends in the group.

13 – The probability is 100%, because no president's picture is on a ten dollar bill. [The portrait is of Alexander Hamilton, who was treasurer, but never president].

14 – To say that whichever two are picked, one is bound to be a chemist, is but another way of saying that no two are physicists. Thus, there was only one physicist in the group and 49 chemists. Hence the chances of picking a chemist at random is 49 out of 50 or 98%.

15 – The probability is 100% for the following reasons: Could there be two of any one color in the group? No, because, say, there were two reds. Then one could pick two reds, one white and one yellow, thus contradicting one of the observations that at least one had to be blue. A similar argument works for each of the other three colors. And so the only possibility is that there were only four flowers in the entire garden–one of each color–and so, if one picks four, then of course one has to be white.

16 – The female writer and the male doctor are both French, hence the other two are German. The German woman cannot be the doctor or writer, both of which are French, nor the lawyer, who is a man. Thus, the German woman is the chemist.

Mr. Schmidt, who is German, cannot be the doctor or writer, who are French, nor the chemist, who is a woman. Thus, Mr. Schmidt is the lawyer. Hence, the German Mr. Schmidt is married to the French writer, and so Mrs. Schmidt is the writer.

17 – The one who is not a secretary is neither blond nor married, hence each of the other two sisters is a blond, married secretary.

18 – Since Arlene is not married and the oldest sister is, then Arlene is not the oldest. Also, Beatrice can't be the oldest, because she is shorter than the youngest, and the oldest is the tallest. Thus Cynthia is the oldest. Also, since Beatrice is shorter than the youngest, she cannot be the youngest, so she is the middle one. This leaves Arlene as the youngest.

19 – There has been much controversy about this problem! Some people favoring Argument 1, and others, Argument 2. At first, Argument 2 might seem the more plausible but it is actually Argument 1 that is correct; the probability is two-thirds, not one-half! The fallacy in Argument 2 is in the last sentence, in which it is said that the probability hasn't changed by adding and removing a black marble. This is not true; the probability *has* changed, since the removed marble was chosen *at random*! By seeing that the remained marble was black, the number of possible cases has been reduced from four to three, and this changes the probability accordingly.

Let's look at it this way: Suppose you do the experiment four times, only using cards instead of marbles. The first two times, put a red card on the table and then a black card to the right of it. The next two times, put a black card on the table and add another back card to the right of it. On the first time, remove the left card; the second time, the right; the third time, the left, and the fourth time, the right. [Thus, you have removed the original card half of the time]. In how many of those three times in which you removed a black card, did a black card remain? Obviously, two.

20 – This is a very different story! Adding a black marble and then *knowingly* removing a black marble effect no change in probability. In this situation, the probability is now one-half that the remaining marble is black.

The four experiments with cards which are relevant to *this* problem are the same as before, except that now, you remove the right card on the first trial, instead of the left. (Thus, you remove a black card all four times). The first two times a red card is left, and the last two times, a black card is left.

21 – Without any information about the color of the removed marble, all four cases (considered in the solution of Problem 16) are equally likely, and in three of them, the remaining marble is black. Thus, the probability is now three-fourths.

To summarize this and the last two problems:

(1) If the removed marble is picked at random and seen to be black, then the

probability (of the remaining marble being black) is two-thirds.

(2) If a black marble is deliberately removed, then the probability is one-half.

(3) Without knowing the color of the removed marble, the probability is three-fourths.

22 – If David is the programmer, we get the following contradiction: David was then born in January (since the programmer was) and is exactly 26 weeks older than Edward, who was born in August. But this is only possible if David was born on January 31 and Edward on August 1 and there is no February 29 in between-in other words, that it was not a leap year. [You can check this with a calendar]. Thus, if David is the programmer, then he was not born on a leap year. On the other hand, if David is the programmer, he was 54 years old in January 1998, hence was born in 1944, which *is* a leap year! Thus, it is contradictory to assume that David is the programmer, and so David must be the engineer.

Chapter 2

Which Lady?

There are four sisters named Teresa, Thelma, Leila and Lenore. Teresa and Thelma, whose names begin with "T", naturally tell the truth at all times, whereas Leila and Lenore, whose names begin with "L" always lie.

1 – You have never seen any of these sisters, but one day you meet one and know that it is one of the four sisters, but don't know which one. She then makes a statement from which you can deduce that she must be Teresa. What statement would do this?

2 – Suppose, instead, she made another statement that convinces you she was Lenore. What statement would work?

3 – Once I met one of the four, and she made a statement from which I could deduce that she must be either Teresa or Lenore, but I couldn't tell which! What statement could that have been?

4 – What statement could she make that would convince you that she is either Thelma, Leila or Lenore, but you couldn't tell which?

5 – What about a statement that would convince you that she is either Thelma, Teresa or Lenore, but you couldn't tell which?

6 – Suppose the lady you meet makes the following two statements:

 (1) Teresa once told someone that I am Leila.

 (2) Thelma has never told anyone that I am Lenore.

Who is she?

7 – What single yes/no question could you ask her to determine which of the four sisters she is?

8 – Is there a single yes/no question you could ask her to determine whether or not she is married?

9 – One day, a man met two of the sisters, one of whom had red hair and the other, black. He fell in love with the red-head on sight, and asked her: "Are you married?" Instead of her answering, the other sister replied: "She is married and she always lies!"

"And who are you?" the man asked the one with black hair. Instead of her answering, the red-head replied: "She is either Teresa or Thelma".

Is the red-head married or not, and is she truthful?

SOLUTIONS

1 – One statement that would work is: "I am not Thelma". Neither Leila nor Lenore could say that, because it is true that neither one is Thelma. Also, Thelma couldn't lie and say that she is not Thelma. Hence, the only one who could say that is Teresa.

2 – A statement that works is: "I am Leila". Neither Teresa or Thelma would falsely claim to be Leila, and Leila wouldn't truthfully claim to be Leila, hence only Lenore could make that claim.

3 – A statement that works is: "I am either Teresa or Leila". Teresa could say that (because she really is Teresa or Leila). Thelma wouldn't make the false statement that she is either Teresa or Leila. Leila wouldn't make the true statement that she is either Teresa or Leila. But Lenore could make the false statement that she is either Teresa or Leila. And so the only ones who could say that are Teresa and Lenore, and there is no way to tell which.

4 – A statement that would work is: "I am Thelma". Thelma could truthfully say that, and both Leila and Lenore could falsely say that. The only one who couldn't say that is Teresa.

5 – A statement that works is: "I am not Lenore". Lenore could falsely say that, and both Teresa and Thelma could truthfully say that. The only one who couldn't say that is Leila. (since she is really *not* Lenore).

6 – The lady's first statement was certainly false, for if it were true, then Teresa did once say that the lady was Lenore, and since Teresa is truthful, the lady would really have to be Lenore, contrary to the assumption that she told the truth. Therefore, the lady is a liar. Hence her second statement is false, and therefore Thelma did once say that the lady is Lenore, hence the lady *is* Lenore (since Thelma is truthful).

7 – There is no such question! The reason is that a yes/no question can have only *two* possible responses, whereas there are *four* possibilities for which sister the lady is. Two responses can distinguish only between two possibilities, not four! [This is a basic principle of computer science].

8 – Certainly! One question that would work is. "Are you either married and Teresa or Thelma, or unmarried and Leila or Lenore? You are asking whether one of the following two alternatives holds:

(1) She is either Teresa or Thelma, and she is married.

(2) She is either Leila or Lenore, and she is not married.

Suppose she answers *yes*. Her answer is either true or false. Suppose it is true. Then one of the alternatives (1) or (2) really holds. It can't be (2) (since she answered truthfully, she can't be Leila or Lenore), hence it is (1), and so she is married. On the other hand, suppose her answer is false. Then neither (1) nor (2) holds, so in particular, (2) doesn't hold. Yet, she must be either Leila or Lenore (since she answered falsely), but since (2) doesn't hold, she must be married.

Thus, if she answers *yes*, then she must be married (regardless of whether she told the truth or lied, and there is no way of determining which). A similar analysis, which we leave to the reader, reveals that if she answers *no*, then she is not married.

9 – Let R be the one with the red hair and B, the one with black hair. They made the following statements:

B: R is married and always lies.

R: B is Teresa or Thelma.

Could R be telling the truth? No, because if she were, then B would really be Teresa or Thelma, hence truthful, hence R would always lie (as B said), and we would have a contradiction. Hence, R lied. Therefore, B is neither Teresa or Thelma, so B lied. Thus, it is *not* true that R always lies *and* is married, yet R always lies, hence R is not married.

In summary, B and R both lied and R is not married, and we can only hope that the man doesn't finally marry R, who always lies!

Chapter 3

Which Witch?

1 – A man was to choose a bride from among three sisters named Alice, Beatrice and Cynthia. It was known that each sister would either always lie or always tell the truth. Also, one and only one of the three sisters was a witch, but it is not given whether the witch lies or is truthful. The three made the following statements to the suitor:

Alice – Cynthia is the witch.

Beatrice – I am not the witch.

Cynthia – At most one of us ever tells the truth!

Now, above all, the man wanted to be sure that the one he marries is *not*, a witch! Which one should he pick? Also, is the witch truthful or not?

2 – According to another version of the above story, the three made the following statements instead:

Alice – I am the witch.

Beatrice – I am the witch.

Cynthia – At least two of us always lie!

If this version is correct, then which one is the witch, and which ones are truthful?

3 – A big robbery occurred in Chicago and three criminals named Mike, Spike and Slug were on trial. It was known that at least one of three was involved, and that no one else was. Mike and Spike were brothers who looked so much alike that they were often confused. Both brothers were quite timid and neither one would ever pull a job without an accomplice. On the other hand, Slug would never trust a partner and always worked alone. The case proved quite difficult until it was

finally discovered that one of the two brothers was seen in a saloon in Houston at the time of the robbery, but it was not certain whether it was Mike or Spike.

Which ones are innocent and which ones are guilty?

4 – On the planet *Blam*, two of the native words are *Tak* and *Bin*; one of them means yes, and the other, *no*. A visitor from our planet once asked a scholar of Blam whether *Tak* meant *no*. The scholar answered correctly, but although he knew English, he absent-mindedly answered in his own tongue. Did he answer *Tak* or *Bin*?

5 – Here is a *metapuzzle* about the famous Island of Knights and Knaves, where knights always tell the truth; knaves always lie, and each native is either a knight or a knave.

A logician once visited this island because of a rumor that there may be gold buried there. He first met a native named *Ark* and asked him: "Is there gold on this island?" Ark replied: "I once claimed there is". This was none too helpful! Then he met a second native *Bork* and asked him if there is gold on the island. Bork's reply was even less helpful: "I never claimed there isn't". Then he met a native *Cag*, whom he knew was acquainted with Ark and Bork, and asked: "Are these two of the same type–both knights or both knaves–or are they of different types?" Cag told him whether or not Ark and Bork were of the same type, but the logician was still in the dark, since he didn't know whether Cag was a knight or a knave. Finally he met a fourth native *Dag*, and asked him: "Are you and Cag of the same type"? Dag answered (*yes* or *no*), and the logician then knew whether or not there was gold an the island.

Is there gold on this island?

SOLUTIONS

1 – It is not possible to tell which one is the witch, but it is possible to tell of one of them that she is definitely *not* the witch. Here is how:

Cynthia either is truthful or she lies. Consider first the case where Cynthia is truthful. Then at most one of the three is truthful, as she claimed, hence Alice and Beatrice both lie. Then Beatrice's statement is false, which means that Beatrice is the witch. This proves that if Cynthia is truthful, then Beatrice is the witch and the witch lies.

Now consider the case that Cynthia lies. Then it is not the case that at most one is truthful, hence Alice and Beatrice must both be truthful, hence Cynthia is the witch, as Alice claimed. And so if Cynthia lies, then she is the witch, and of course the witch then lies.

This leaves Alice in the clear, as far as being a witch is concerned, so the suitor, to be safe, should pick Alice. Whether Alice is truthful or not cannot be determined (and hopefully will be in the course of the marriage).

2 – This one is simpler: If Cynthia lied, then at most one lies, hence she is the only liar, hence Alice and Beatrice both spoke the truth, which is impossible, since only one of the three is a witch. Therefore, Cynthia is truthful. Since her statement was true, then Alice and Beatrice both lied, hence neither one is really the witch, so it must be Cynthia. So according to this version, Cynthia is the witch and the only truthful one of the three!

3 - The brother who was in Houston is obviously innocent. The other brother must also be innocent, since he didn't work alone, nor with Slug, who only works by himself. So Slug is the only guilty one.

4 – The question asked was whether *Tak* means *no*. If Tak does mean *no*, then the correct English answer would be *yes*, hence the correct native answer would be *Bin*. On the other hand, if Tak means *yes*, then the correct English answer to the question would be *no*, hence the correct native answer would again be *Bin* (which means *no*). So in either case, the correct native answer to the question is *Bin*.

5 – This is called a *metapuzzle* because we are not told what either Cag or Dag said, and can only solve the problem by knowing that the logician was able to solve it.

First we must see that after Dag answered, the logician knew whether or not Ark and Bork were of the same type. Well, after Dag answered, the logician knew whether Cag was a knight or a knave. For suppose Dag answered *yes*. Then Cag would have to be a knight, because no native would claim to be the same type as a knave (no knight would lie and say he is, and no knave would truthfully admit that he is). On the other hand, if Dag said *no*–if he denied being the same type as Cag–then Cag would have to be a knave. And so, after Dag answered, the logician knew whether Cag was a knight or a knave. Once he knew that, then he knew whether or not to abide by Cag's answer, and hence then knew whether or not Ark and Bork were of the same type.

Now, the crucial point is that if Ark and Bork are of different types, then there is no way of knowing whether there is gold on the island, for it could be that Ark is a knight and there is gold, or it could be that Ark is a knave, Bork is a knight and there might or might not be gold. And so, if the logician had found out that Ark and Bork were of different types, then he couldn't have known whether there was gold on the island. On the other hand, suppose that he found out that Ark and Bork were of the same type. Well, suppose they are both knights. Then Ark is a knight, so he really did once claim there was gold (as he said), and being a knight, there really must be gold. Thus, if they are both knights, then there is gold. What if they are both knaves? In that case, Bork is a knave, hence his statement was false, hence he once *did* claim that there was no gold, and being a knave, there really is gold.

This proves that if Ark and Bork are either both knights or both knaves, there must be gold, whereas if they are of different types, there may or may not be gold, and there is no way to determine which. But we are given that the logician *could* determine which, and so from Cag's and Dag's statements, he must have known that Ark and Bork *were* of the same type, and hence there must be gold on the island. (Whether Ark and Bork were both knights or both knaves cannot be determined, nor can we know the type of Cag or of Dag, though the logician knew the type of Cag, but not of Dag.)

Chapter 4

Which Island?

Amin and Begone are two neighboring islands in some far-off ocean. The natives of Amin (the Aminians) and the natives of Begone (the Begonians) sometimes visit each other's island. The strange thing is that whenever the natives are on their home island, they tell the truth, but whenever they are on the other island, they lie.

1 – Suppose that you land on one of the islands, but don't know which one it is. You meet a native who you know is either Aminian or Begonian, but you don't know which. You are allowed to ask him only one yes/no question, and your purpose is to find out which island you are in. What question would you ask?

2 – Suppose that instead of wanting to know which island you are on, you wanted to know whether the native you met is Aminian or Begonian. Again, you are allowed only one yes/no question. What question would you ask?

3 – Suppose that you don't care which island you are on, nor whether the one you are speaking to is Aminian or Begonian, but you want to know whether he is now on his home island. What question would work?

4 – Suppose all you want to do is make him answer *yes*. What question would work?

5 – Suppose you meet a native who says: "When I am on Amin, I can claim that I am on Begone". Which island are you on?

6 – Suppose the native says: "When I am on Amin, I claim that I am Begonian. I am really Aminian".

From this, it is possible to deduce *both* which island you are on, and which island that native is from. What is the solution?

7 – [A Metapuzzle] – One day a logician landed on one of the islands, but didn't know which one. He then met two natives, but didn't know whether they were from the same island or not. One of them said: "We are from different islands". Then the logician asked: "What island are we now on?" One of the two replied: "Amin". The logician then knew which island he was on. Which island was it?

8 – Here is a more complex one: A visitor once came to one of the two islands and met two natives named *Auk* and *Bog*. They made the following statements:

  Auk: At least one of us is Aminian.

  Bog: Auk is Begonian.

  On which island did this take place, and what are Auk and Bog?

SOLUTIONS

1 – Many questions would work, but a particularly simple one is: "Are you Aminian"? Suppose he answers "yes". He is either now truthful or lying. Suppose he is telling the truth. Then he is really Aminian, as he claims, but also must now be on Amin (in order to be truthful), and so you are now on Amin. But suppose he is lying? Then, he is not really Aminian, as he claimed, hence he is Begonian. Since he is lying and is Begonian, he must now be on Amin, and so again you are now on Amin. This proves that if he answers *yes*, then regardless of whether he is telling the truth or lying, you are now on Amin. By a similar analysis, which we leave to the reader, it can be seen that if he answers *no*, then you must now be on Begone.

2 – To find out where he is from, ask him "Are you now on Amin?" By a fairly similar analysis to that of the last problem, if he answers *yes* then he is Aminian, and if he answer *no*, then he must be Begonian.

    There is a pretty symmetry between this and the last problem: To find out whether you are on Amin, you ask: "Are you Aminian?" but to find out if he is Aminian, you ask: "Are you now on Amin?"

3 – You need merely ask: "Does two plus two equal four?"

4 – Just ask: "Are you now on your home island?" You are bound to be answered *yes* (because if he is on his home island, he will truthfully answer *yes*, and if he is *not*, he will falsely answer *yes*).

5 – If he is Begonian, then his claim is true, which means he is now on Begone. If he is Aminian, his claim is false (because he couldn't then falsely claim to be on Begone), hence he must now be on Begone. So in either case, you are now on Begone.

6 – It is impossible that both statements are true, because if he is really Aminian (as he claimed in his second statement), then when on Amin, he wouldn't make the false claim that he is Begonian. Therefore, both his statements were lies. Since his second statement was false, then he is not really Aminian, but Begonian. Since he is Begonian and lying, then he must be on Amin. And so you are on Amin and talking to a Begonian.

7 – Let A be the person who said: "We are from different islands" and B be the other one. If A's statement was true, then A was on his home island, hence B was off his home island (as A's statement implied). On the other hand, if A's statement was false, then A and B were actually from the same island, but A was off his home island (having made a false statement), hence B was also off his home island. Thus in either case, B was off his home island, hence in the lying state. There is no way of knowing whether A was in the lying state or not. Now, we are not told whether it was A or B who answered the logician's question, but if it were A, the logician could have had no way of knowing where he was. But the logician *did* know, hence it must have been B who answered "Amin", and the logician knew he was lying (as we have shown), and hence knew that he was really on Begone.

8 – Could Auk be Begonian? Well, suppose he were, then we get the following contradiction: The island they are on is either Begone or Amin. Suppose it is Begone. Then, Auk, being Begonian, told the truth, hence at least one of the two is really Aminian, hence it must be Bog. But how could Bog, being Aminian on the island of Begone, have made the true statement that Auk is Begonian? Thus the island can't be Begone (assuming Auk is Begonian), but it also can't be Amin either, for suppose it is Amin. Then Auk, being Begonian on Amin must have lied, which means that neither one is really Aminian, hence both are Begonian. But then, how could Bog, being Begonian and on Amin, have truthfully said that Auk is Begonian? Thus the assumption that Auk is Begonian leads to a contradiction (regardless of whether the island is Begone or Amin), and so Auk is really Aminian. Since Auk is Aminian, then it is true that at least one of the two is Aminian, thus Auk told the truth. Since he told the truth and is Aminian, the island they are on must be Amin. Also Bog lied in saying that Auk was Begonian, hence Bog must be Begonian (having lied on Amin).

In summary, the island is Amin; Auk is Aminian and Bog is Begonian. Auk told the truth and Bog lied.

Chapter 5

McGregor's Arithmetic Tricks

For a change of pace, we now consider some arithmetical recreations in this and the next two chapters.

George, a high school student, walked into McGregor's shop one day to buy a billiard ball and asked the price. Now, Mr. McGregor was in a generous mood at the time, and having a general interest in mathematical education of the young, said: "I'll tell you what. I'll give you a series of puzzles, and if you succeed in solving them, then I'll give you a billiard ball free. Fair enough?"

"I'm game to try", said George.

1 – "Good", said McGregor. "Let's start with a simple one. I recently counted the number of pennies I now have in this store and realized that I could divide them into eight even piles, or seven even piles, or six even piles, but if I had any fewer pennies, then I couldn't do these three things. Can you tell me how many pennies I have?"

George got this one easily.

2 – "Good", said McGregor. "Now let me tell you that yesterday I went into a store and spent half of what I had. Then I took a subway for a dollar, and then went into another store and spent two-thirds of what I then had. Then I left and bought a newspaper for 50 cents. Then I went into another store and spent four-fifths of what I then had. When I walked out, I saw a beggar and gave him a quarter. I then had left only one dollar, with which I took a subway home. Can you tell me how much I started out with?"

3 – "Speaking of cards", said McGregor, "suppose that you arrange the thirteen cards of one suit from left to right in the order A, 2, 3, 4, 5, 6, 7, 8, 9, 10, J, Q, K. You put your finger on A and count "one". Then you put your finger on 2 and

count "two" and you keep going right until you come to K, counting "thirteen". Then you reverse direction, and count Q as "fourteen", J as "fifteen" until you come back to A, counting "twenty-five". Then you reverse again and go right, and you thus keep going back and forth until you count "one million" and then stop. On what card will you then be?"

"Good gracious!" said George. "I haven't all day!"

"Oh, you don't actually have to count to a million to get the answer", said McGregor. "There is a simple trick to solve the problem quite quickly".

George thought for a moment and got the trick.

What is the answer?

4 – "Very good!" said McGregor. "Now for a more interesting one. Let me tell you about *special numbers*".

"What are they?" asked George.

At this point McGregor handed George a pencil and a pad of paper. "Turn around", said McGregor, "and write down any 3-digit number. Don't let me see what you write".

George wrote down 349.

"Now write down that number followed by itself".

George then wrote 349349.

"Now divide what you have by 7".

George did so and got 49907.

"Now divide what you have by eleven".

George did so and got 4537.

"Now divide what you have by thirteen".

George did so, and to his surprise got back his original number 349. "That's curious!" said George, "I got back the same number I started with!"

"Ah, that's because you started with a *special* number!" said McGregor.

"You still haven't told me what a special number is", protested George.

"By a *special* number", replied McGregor, "I mean any 3-digit number such that when repeated and then successively divided by seven, eleven and thirteen, gives you back the number you started with. As you can see, you happened to start with a special number".

"Are there others? asked George.

"Yes", said McGregor, "and your problem is to tell me how many there are".

"Oh, no!" said George. "You mean I have to test every number from 100 to 999?"

"No", said McGregor. "There is again a simple, but clever, method of solving this without all that work".

George had to work quite hard on that one, but he finally saw the trick and laughed at its beautiful simplicity.

How many special numbers are there?

5 – "Very clever", said McGregor. "You have real mathematical talent. Now, for your last problem, you must tell me how many billiard balls I have in this store. If you do that, then, as I have promised, I will present you with a billiard ball.

"My billiard balls are all of the same weight, and collectively weigh eighty-seven pounds and eleven ounces. Each ball weighs a whole number of ounces; no fractions of an ounce are involved".

"I can't believe you have given me enough information", said George. "For all I know, you might have just one giant billiard ball weighing eighty-seven pounds and eleven ounces!"

"I never thought of that" laughed McGregor, "but I assure you that I have more than one billiard ball".

"Then there is another possibility", said George after a short calculation. "Eighty-seven pounds and eleven ounces is 1403 ounces, and for all I know, you might have 1403 tiny billiard balls each weighing one ounce!"

"What an idea!" said McGregor. "No, I can assure you, each ball weighs more than an ounce".

George then went to work with pencil and paper. After a while, he said: "There is still more than one possibility!"

"That's right", McGregor suddenly realized, "so let me add that each ball weighs more than two pounds.

"Oh, good!" said George. "Now I know how many billiard balls you have".

How many billiard balls were there?

SOLUTIONS

1 – The number 8x7x6 is, of course, divisible by 8, 7 and 6, but is not the smallest such number, which is 8x7x3, which is 168. And so, he could make eight piles of 21 each, or seven of 24 each, or six piles of 28 each.

2 – We solve this by working the problem backwards. Before giving 25 cents to the beggar, McGregor had $1.25. Then, before his third store purchase, he must have had $6.25. Then, before he bought the newspaper, he had $6.75. Then, before he made his second store purchase, he had $20.25. Then, before the subway, he had $21.25. Then, before his first store purchase, he had $42.50.

3 – By a *cycle*, let us mean the 24 steps consisting of going from A to K and back to 2. Now, 1,000,000 divided by 24 is 41,666 with a remainder of 16. And so after 41,666 cycles, you will have made 999,984 steps and be on 2 going left, with 16 more steps to go. After these 16 steps, you will land on 10.

4 – Following a 3-digit number by itself is tantamount to multiplying it by 1001! Also, 7x11x13=1001. And so going through the process described is tantamount to first multiplying the number by 1001 and then dividing the result by 1001, which of course gives you back the original number! Thus *every* 3-digit number is

"special", and so there is nothing very special about "special" numbers. Of course, there are 900 3-digit numbers. (999-99), and so the answer is 900.

5 – The balls collectively weigh 1,403 ounces, and 1,403 is the product of the prime numbers 23 and 61. And so there are either 23 balls each weighing 61 oz. or 61 balls weighing 23 oz., but the latter possibility is ruled out, since we were told that each ball weighs more than 32 oz. And so there are 23 balls each weighing 61 ounces.

Chapter 6

Ask Eldon White

1 – Eldon White decided one day to return a bicycle he had borrowed from a friend. He rode the bicycle to his friend's house at the rate of 9 miles an hour. He then walked back home at the rate of 3 miles an hour. Altogether he was gone 8 hours. How far is his house from that of his friend?

2 – Eldon has 4 dogs. One day he put out a bowl of dog biscuits. The eldest dog came first and ate half of the biscuits plus one more. Then the next dog came and ate half of what he found plus one more. Then the next one came and ate half of what she found plus one more. Then the little one came and ate half of what she found and one more, and that finished up the biscuits. How many biscuits were originally in the bowl?

3 – Eldon once bought a very remarkable plant which, on the first day, increased its height by 1/2, then on the second day by 1/3, then on the third day by 1/4, and so on. How many days did it take to grow to 100 times its original height?

4 – Eldon has 4 children. The youngest one, Betty, is 9 years old. Then there are the twin boys, Arthur and Robert. Then there is Laura, the eldest, whose age is equal to the combined ages of Betty and Arthur. Also, the combined ages of the twins are the same as the combined ages of the youngest and eldest. How old is each?

5 – "Give us a puzzle, Dad", said Arthur one evening. "Very well", said Eldon, who then went into another room and returned about ten minutes later. He then put 3 cards face down on the table. On the back of each was written a statement. Eldon explained that if the card was red, the sentence written on it was true, but if the card was black, the sentence written on it was false. Here are the 3 backs.

A	B	C
Exactly one of these 3 cards is black	Exactly 2 of these 3 cards are black	All three of these cards are black

What color is each of the three cards?

6 – "How about a riddle?" asked Robert.

"Very well", said Eldon. "What is it that is larger than the universe; the dead eat it, and if the living eat it they die?

What is it?

SOLUTIONS

1 – Let x be the distance between the two houses. Then Eldon rode for x/9 hours and walked for x/3 hours, and so x/9+x/3=8. This makes x=18.

2 – This problem is best solved by working it backwards. How many biscuits did the fourth dog find? Well, dividing that number by 2 and subtracting 1, we get 0. Reversing the procedure adding 1 to 0 and then multiplying by 2, we get 2. Thus, the last dog found 2 biscuits. Adding 1 and multiplying by 2, we get 6, the number of biscuits found by the third dog. Adding 1 and multiplying by 2, we get 14, the amount found by the second dog. Finally, adding 1 and multiplying by 2, we get 30, the amount originally in the bowl.

3 – The plant was originally, say, one unit tall (the length of a unit really doesn't matter). After one day, the plant was 1 1/2 units tall. On the next day, it gained 1/3 of 1 1/2, which is 1/2, hence was then 2 units tall. The next day it gained 1/4 of 2, which is again 1/2. And so, the plant actually gained 1/2 unit each day. After 198 days it gained 99 units, and was then 100 times as tall as on the first day. Thus the answer is 198 days.

4 – Let x be the age of each twin and y be the age of Laura. Then y=x+9 and y+9=2x. This makes x=18 and y=27. Thus, the twins are 18 years of age and Laura is 27.

5 – C is obviously black, because if it were red, the sentence on its back would be true, which would mean that all three are black, which is a contradiction. Thus C is black. It then follows that the sentence it bears is false, hence not all three are black and so at least one of the cards is red. Could it be A? No, because if A were red, its sentence would be true, which means that only one card is black, hence B would have to be red, which would mean that B's sentence was true, which would mean that exactly two are black, contrary to the fact that only one is black (assuming A is red). Thus A must be black, and so it is B that is red. Thus A and C are black and B is red.

6 – Nothing is larger than the universe. The dead eat nothing. If the living eat nothing, they die. So the answer is *nothing*.

Chapter 7

Al, the Chemist

1 – Al, the chemist, was not an alchemist, as his name might suggest. Anyway, one day, he partially filled a container with some concoction or other. He knew the volume of fluid in the container, as well as the volume of empty space and realized that 2/3 of the former was equal to 4/5 of the latter. Was the container then less than half full, more than half full, or exactly half full?

2 – On another occasion. Al had 100 cc's of an alcohol and water mixture, 5% of which was alcohol. He wished to reduce the concentration to 4% alcohol. How much water must he add?

3 – On still another occasion, Al had two beakers, one of which contained 10 ounces of water and the other, 10 ounces of wine. He poured 3 ounces of water into the wine beaker, stirred the mixture, and then poured 3 ounces of the mixture back into the water beaker. Which was then more, the amount of water in the wine beaker, or the amount of wine in the water beaker?

4 – Now suppose that Al continued the process, pouring 3 ounces back and forth from one beaker to the other. How many pourings would be necessary to reach a point of equilibrium–a point at which the concentration of wine in both beakers would be the same?

SOLUTIONS

1 – Let x be the amount of fluid in the container and y be the amount the container would hold if full. Then, the volume of empty space is y-x, and so $\frac{2x}{3} = \frac{4(y-x)}{5}$. This makes x=6/11y, and so the container is 6/11 full, which is more than half.

2 – Initially, there are 5 cc of alcohol and 95 cc of water (and thus 100 cc of mixture). Let x be the amount of water, the percentage of alcohol to the mixture is 5/100+x. And so, we want x to be such that 5/100+x=4/100. This makes x=25.

3 – Since the volume of fluid in the water beaker was the same after the two pourings as before (10 oz.), then whatever water is missing, must be replaced by the same volume of wine. Therefore, the amount of wine in the water beaker is *the same* as the amount of water in the wine beaker.

4 – If water and wine were absolutely homogeneous, the answer would be that no finite number of pourings could suffice, because to begin with, the water beaker is obviously weaker (in wine) than the wine beaker, and at every stage, if the pouring is from the water beaker to the wine beaker, the water beaker still remains weaker, and if the pouring is from the wine beaker to the water beaker, the water beaker is still weaker in wine concentration. Thus no finite number of pourings can suffice.

Theoretically, that answer would be fine, if we were dealing with homogeneous fluids, but in actual fact, both water and wine consist of *discreet* molecules, and so from a practical point of view, the above answer is not valid. From a physical, as opposed to a purely mathematical, point of view, an equilibrium is possible after a finite number of pourings, but the number of necessary pourings cannot be predicted, since there are chance factors involved. According to one physicist, after 47 back and forth pourings, the chances of equilibrium is greater than fifty percent.

Chapter 8

Sane or Mad?

Now we return to puzzles in logical reasoning.

In a certain country, half the inhabitants are totally mad and believe all true propositions to be false and all false propositions to be true. In other words, *all* their beliefs are wrong! The other half of the inhabitants are completely sane and totally accurate in all their beliefs. All the inhabitants–mad and sane–are completely honest and always state what they really believe.

As to be expected, there are many psychiatrists in this country, but to complicate matters, some of the psychiatrists are also mad! This leads to some curious situations, as the reader will soon see.

1 – A certain patient believed that he and his psychiatrist were not both sane. Which one, if either, is sane and which one, if either, is mad?

2 – Another patient believed that he and his psychiatrist were both mad. Is the solution the same as that of the last problem?

3 – If an inhabitant says: "I believe that I am sane", is he or she necessarily sane?

4 – Does every inhabitant of this country believe that he believes that he is sane?

5 – One day a certain Dr. Schultz was being interviewed and said: "I am not a sane psychiatrist". Can his sanity be determined? Can it be determined whether or not he is a psychiatrist?

6 – Is it possible for an inhabitant to say: "I am a mad psychiatrist"?

7 – (a) – A patient once complained to his psychiatrist: "You don't believe I'm sane! You believe I am mad!"

Can the sanity of either the patient or the psychiatrist be determined from this?

(b) – Then the psychiatrist replied: "How can you say that? I never believed that you are mad!"

What can now be deduced about each?

8 – This country has a king, but it is not given whether this king is sane or mad. Suppose that you visit the country and meet an inhabitant whom you suspect might be the king. You wish to find out if he is, but you are allowed to ask him only one question answerable by *yes* or *no*. What question would you ask?

9 – One day, someone asked an inhabitant named *Bog*: "Didn't you once claim that you are the king?" Bog replied: "Certainly not. I am not the king!"

Can it be determine whether or not Bog is the king? Can it be determined whether he is sane or mad?

10 – On another occasion, an inhabitant named *Org* was asked whether he was the king. He gave the quizzical reply: "If I am sane, then I am the king".

Can it be determine whether he is sane? Can it be determine whether or not he is the king?

SOLUTIONS

1 – If the patient were mad, then it would be true that he and his psychiatrist were not both sane, hence the mad patient would have a true belief, which is not possible. Therefore, the patient must be sane. Being sane, his belief is true, and so he and his psychiatrist are not both sane, but he is, so his psychiatrist isn't. Thus the patient is sane and the psychiatrist is mad.

2 – No, this is a very different situation! In the first problem, the patient believed that at least one of the two (patient and psychiatrist) was mad, whereas in the present problem, the patient believes that *both* are mad. Now, a sane patient couldn't possibly believe that he and the psychiatrist were both made, hence the patient must be mad. Being mad, his belief is false, hence the two are not both mad, and so the psychiatrist is sane. Thus the solution to this problem is the very opposite of the solution to the first problem. In this problem the patient is mad and the psychiatrist is sane.

3 – Many readers will be surprised that the person is necessarily sane! Every inhabitant of this country, sane or mad, believes himself to be sane, hence when the inhabitant said: "I believe I am sane", the statement was true, hence the inhabitant must be sane.

4 – No, only the sane ones do. Let me explain: One of the peculiarities of the mad people of this country is that whenever a mad inhabitant believes something, he doesn't believe that he believes it; in fact, he believes that he *doesn't* believe it!

The reason is that if he believes something, then since it is *true* that he believes it, he cannot believe the true fact that he believes it–rather, he believes the *false* proposition that he doesn't believe it. And so, although every mad inhabitant believes he is sane, he doesn't believe that he believes he is sane.

5 – If Dr. Schultz were mad, it would be true that he is not a sane psychiatrist, hence he would have said something true, which a mad person cannot do. Hence, he is sane, and furthermore, he is not a sane psychiatrist (as he correctly claimed). Thus, Dr. Schultz is sane, but not a psychiatrist.

6 – Yes, a mad non-psychiatrist could make a false statement that he is a mad psychiatrist.

7 – (a) – The patient is either sane or mad. Suppose he is sane. Then it is true, as he said, that the psychiatrist believes he is mad, and so in this case, the psychiatrist must be mad. On the other hand, suppose that the patient is mad. Then his statement that the psychiatrist doesn't believe he is sane is false, which means that the psychiatrist *does* believe that he is sane, so again the psychiatrist has a false belief, hence is mad. So regardless of whether the patient is sane or mad, the psychiatrist must be mad. At this stage, it cannot be determined whether the patient is mad or sane.

(b) – But now the psychiatrist said that he never said that the patient was mad. Being mad, the psychiatrist made a false statement, which means that he once *did* believe that the patient was mad, but since that earlier belief was also false, the patient is really sane.

8 – All you need ask him is: "Do you *believe* you are the king?" Suppose he answers *yes*. If he is sane, then he really does believe he is the king, as he said, and being sane, his belief is correct, and so he is then the king. But suppose he is mad? This is the more interesting case: If he is mad, then his *yes* answer was incorrect, hence he doesn't really believe that he is the king (he only mistakenly affirmed that he does), and since he doesn't believe he is the king, then he really must be the king (because if he weren't, he would falsely believe that he was). And so, if he is mad and answers *yes*, he is again the king. This proves that if he answers *yes*, then regardless of whether he is sane or mad, he must be the king. Mind you, if an inhabitant believes he is the king, he is not necessarily the king (he might be mad), but if he *says* that he believes he is the king, then he believes that he believes that he is the king, hence he really must be the king.

A similar analysis, left to the reader, reveals that if you get the answer *no* to your question, then he is not the king (regardless of whether he is mad or sane).

9 – If Bog were mad, we get the following contradiction: Suppose Bog is mad. Then his first reply "Certainly not" was false, which means that he did once claim to be the king, hence he is not really the king, but then his second statement "I am not the king" would be true, which is not possible for a mad inhabitant. Thus

Bog must really be sane. And so Bog is sane, he is not the king, and never claimed he was.

10 – Org claimed that if he is sane, then he is the king. Let us see if he is right; Suppose he is sane. Then his statement is true, from which follows that he must be the king. This doesn't prove that he *is* the king; all it proves is that *if* he is sane, then he is the king. And so we now know that if he is sane, then he is the king. Well, he said just that, hence what he said was true, and so he must be sane! It then further follows that he must be the king. And so he is both sane and the king.

Some readers will say: "But you haven't considered the case that he is *not* sane!" My reply is that we don't need to consider that case, because I have already proved that he *is* sane (he made the true statement that if he is sane, then he is the king).

Some of you are still unconvinced! I expect a storm of protest saying that my argument is circular. I can assure you it is not!

Chapter 9

The Strange Case of McSnurd

1 – **Introducing McSnurd** — Cornelius McSnurd is a most unusual individual! On Mondays and Tuesdays he is completely sane and knows which propositions are true and which ones are false, but on Wednesdays and Thursdays, he is totally mad and believes all true propositions to be false and all false propositions to be true. On these two days he is totally deluded in all his judgments. To complicate matters, on Mondays and Wednesdays he is completely truthful and honestly tells you what he really believes, but on Tuesdays and Thursdays, he is a complete liar and always states the opposite of what he actually believes. Thus on Mondays and Thursdays, he will make only true statements and answer all questions correctly, but on Tuesdays and Wednesdays he makes only false statements (either out of malice or delusion) and answers all questions falsely. For example, suppose you ask him whether two plus two is four. On Mondays he is both sane and truthful, hence will answer *yes*. On Thursdays he is both insane and mendacious, hence on this day, he will falsely believe that two plus two *doesn't* equal four and then he will lie and say that it *does*, and so he will also answer *yes*! On Tuesdays he is sane and mendacious, hence he will believe that two plus two equals four and then lie and say *no*. On Wednesdays he is both deluded and truthful, hence he will believe that two plus two doesn't equal four, and true to his belief, he will answer *no*.

(Apropos of McSnurd's behavior on Thursdays, there is a true incident of a schizophrenic whom the doctors were thinking of releasing from a mental institution, but they decided to give him a test under a lie detector. They asked him: "Are you Napoleon?" He answered, "No". The machine showed he was lying!)

Coming back to McSnurd, I forgot to tell you that on Fridays, Saturdays and Sundays, he sleeps all day and never says a word. Now, suppose you meet McSnurd on one of the first four days of the week, but for some odd reason, you are not sure which day it is. What is the minimum number of yes-no questions you could ask him to find out, and what questions would you ask?

2 – On Mondays and Tuesdays McSnurd goes hunting, but on Wednesdays and Thursdays, he goes fishing. Suppose you meet him one morning of the first four days of the week and without knowing what day it is, what single yes-no question could you ask him to reveal whether he will go hunting or fishing on this day?

3 – What question is such that you'll get the same answer on each of the first four days?

4 – Suppose you meet McSnurd on one of the first four days, but do not remember which of the four days it is. You want to find out whether he is truthful or not on this day, but may ask him only one question answerable by *yes* or *no*. What question would you ask?

5 – If you should ask him: "Is your name McSnurd?" then on Mondays and Thursdays you'll get the correct answer *yes*, and on Tuesdays and Wednesdays you'll get the wrong answer *no*. Now, suppose you ask instead: "Do you *believe* that your name is McSnurd?" What answers will you get on each of the four days? (The solution is quite tricky and illustrates a basic principle that will be needed in the next two problems.)

6 – One day an absent-minded logician who had forgotten the day of the week asked McSnurd what day it was. McSnurd answered (he named one of the first four days), and the logician then knew what day it was.

Later in the day, a second absent-minded logician asked McSnurd what day he *believed* it was. McSnurd answered (again he named one of the first four days) and this logician then knew what day it was. What day was it?

7 – **The Mysterious Writing** — One day an experimenter wrote down on a piece of paper the name of one of the first four days of the week. On each of these four days, he showed the slip to McSnurd and asked: "Is today this day?" On three days McSnurd answered *no* and on one day he answered *yes*. If instead, the experimenter had asked: "Do you *believe* that today is this day?" then again he would have received three *no* answers and one *yes* answer. What day was written on the paper?

SOLUTIONS

1 – One question couldn't possibly work, because you can get only two possible responses–*yes* or *no*–, which cannot possibly tell you which of the *four* possibilities holds. With two questions, however, there are many possible solutions. One obvious one is to first ask whether two plus two equals four. If he answers *yes*, you will know that he answers questions correctly and that it is either Monday or Thursday. Then you ask him whether it is Monday and abide by his answer. If he answers *no* to your first question, then you will know that he answers questions incorrectly

and that it is either Tuesday or Wednesday. Then you ask him if it is Tuesday and then disbelieve his answer.

2 – A question that works is: "Is today either Monday or Wednesday?" On Monday he will correctly answer *yes*, on Tuesday he will incorrectly answer *yes*, on Wednesday he will incorrectly answer *no* and on Thursday he will correctly answer *no*. So if he answers *yes*, it is his hunting day (Monday or Tuesday), and if he answers *no*, it is his fishing day.

3 – A question that works is: "Is today either Monday or Thursday?" On each of the four days you'll get the answer *yes*.

4 – Ask: "Is today Monday or Tuesday?" If he answers *yes*, he is truthful (but not necessarily correct), and if he answers *no*, he is lying (but his answer might be correct). I leave the proof to the reader.

5 – If asked whether he *believes* his name is McSnurd, then it is obvious that on Monday he will answer *yes* and on Tuesday he will answer *no* (since on those two days he *does* believe that his name is McSnurd, and he is truthful on Monday and lies on Tuesday). Now, on Wednesday he is insane and doesn't believe that his name is McSnurd and since he is truthful on that day, it may be tempting to think that he will answer *no*, but this is wrong! Since on Wednesday it is *false* that he believes he is McSnurd, and since on that day he believes *all* false propositions, then he wrongly believes that he *does* believe that his name is McSnurd, and true to this wrong belief, he answer *yes*. The whole point now is to realize that on days when he is insane (Wednesdays and Thursdays) whenever he believes something, he believes that he *doesn't* believe it, and whenever he doesn't believe something, he believes that he *does* believe it. (This principle is important for the next two problems as well). And so on Wednesdays and Thursdays he (wrongly) believes that he believes that his name is McSnurd, and so on Wednesday he will truthfully answer *yes* and on Thursday, he will lie and say *no*.

6 – We first note the following four facts:

(1) On Mondays, Tuesdays and Wednesdays, McSnurd could claim that it was Monday (rightly so on Mondays and wrongly on Tuesdays and Wednesdays).

(2) Only on Wednesday could he claim it was Tuesday (wrongly, of course).

(3) Only on Tuesdays could he claim it was Wednesday (again wrongly).

(4) On Tuesdays, Wednesdays and Thursdays he could claim it was Thursday (wrongly so on Tuesdays and Wednesdays, and rightly so on Thursdays).

Therefore, if the first logician had been told that it was Monday, he couldn't have known whether it was Monday, Tuesday or Wednesday, and if he had been told that it was Thursday, he couldn't have known whether it was Tuesday, Wednesday or Thursday, but he *did* know, hence it must be that either he was told that it was Wednesday and then knew that it was Tuesday, or he was told that it was Tuesday and then knew that it was Wednesday. So now you and I know that the day must be Tuesday or Wednesday, but we don't know which (although the first logician did).

As for the second logician's question, we note the following four facts (using the principle that on Wednesdays and Thursdays, whatever McSnurd believes, he believes he *doesn't* believe, and whatever he doesn't believe, he believes he *does* believe).

(1) On Mondays, he obviously could claim to believe it was Monday, and couldn't claim to believe it was any other day.

(2) On Tuesdays, he believes it is Tuesday, and then lies about his belief, hence he could give any of the answers *Monday, Wednesday* or *Thursday*.

(3) On Wednesday, he is crazy and doesn't believe that it is Wednesday, so he believes that he *does* believe it is Wednesday, and true to his belief, he would answer "Wednesday".

(4) On Thursday, he would believe it was not Thursday, hence would believe that he *did* believe it was Thursday, but then he would lie about his belief, so instead of answering *Thursday*, he would either answer *Monday, Tuesday* or *Wednesday*.

Thus, the answer *Monday* would indicate that the day was either Monday, Tuesday or Thursday, but there would be no way to tell which. A *Tuesday* answer would indicate that it must be Thursday. If he answers *Wednesday*, then it could be any of the three days–Tuesday, Wednesday or Thursday. But a *Thursday* answer could be given only on Tuesday. Since the second logician could tell the day from the answer, then the day must be either Tuesday or Thursday, and since we have already seen that the day must be either Tuesday or Wednesday, then the day must be Tuesday.

7 – The experimenter showed the word he had written and asked on each of the first four days whether today was the day that was written. If the written word were *Monday*, he would have received three *yes* answers (on Monday, Tuesday and Wednesday), so the word couldn't be *Monday*. If the word was *Thursday*, he would also have received three *yes* answers (on Tuesday, Wednesday and Thursday), so the word could be *Thursday*. But if the word was *Tuesday*, he would have received only one *yes* answer (on Wednesday), and if the word were *Wednesday*, he again would have received only one *yes* answer (on Tuesday). Therefore, the word was either *Tuesday* or *Wednesday*.

Now, suppose the experimenter had instead asked if McSnurd *believed* that this was the day. If the word were *Wednesday*, he would have received only one *no* answer (on Monday, because on Tuesday McSnurd would know it wasn't Wednesday and lie, but on Wednesday, he wouldn't believe it was Wednesday, hence he would believe that he *did* believe it was Wednesday, and true to his belief, he would say *yes*, and on Thursday, he would believe that it was Wednesday, hence wouldn't believe that he believed that it was Wednesday, and then he would lie and say *yes*). Therefore, the word must be *Tuesday* (and only on Thursday would the answer be *yes*).

Chapter 10

The Knight-Knave Disease

As many of my readers know by now, on the Island of Knights and Knaves, knights make only true statements and knaves make only false ones and each inhabitant is either a knight or a knave. At least, that is how things *normally* are on this island. But one day a strange epidemic known as the *knight-knave disease* struck about half the inhabitants, which made them reverse their roles! Thus sick knights lied, whereas healthy knights told the truth. Sick knaves told the truth, whereas healthy knaves lied. And so in general, when an inhabitant lied, there was no apparent way of knowing whether he was a sick knight or a healthy knave, and if a person told the truth, he could be either a healthy knight or a sick knave. I say, *in general*, there was no way to tell, but there are exceptions, as some of the following problems will reveal.

1 – One day a native made a statement from which it follows that he must be a sick knight. What statement could it have been?

2 – What statement could be made only by a sick knave?

3 – One day a native made a statement from which it can be deduced that he must be either sick, or a knight, or maybe both, but there is no way to tell which. What statement would work?

4 – Suppose, instead, he had made a statement from which can be deduced that he is either sick, or a knight, but *not* both! What statement would work?

5 – On another occasion a native made a statement from which it can be deduced that he must be a knave, but he could be either a sick or a healthy one. What statement would work?

6 – What statement could be made by either a healthy knight or sick knight, a healthy knave or a sick knave?

7 – One day two inhabitants named Aaron and Bowaine made the following statements about each other.

Aaron: Bowaine is a knight.

Bowaine: Aaron is a knave.

Aaron: Bowaine is sick.

Bowaine: Aaron is healthy.

Which of the four types is each of them?

8 – One day a crime was committed and the court knew that it was done by a sick knight. Two men named Archie and Benedict were on trial, and the presiding judge was Inspector Craig of Scotland Yard, who happened to be visiting the island at the time. Here is a transcript of the trial:

Craig: Are either of you sick knights?

Archie: Both of us are.

Craig: Are either of you healthy knaves?

Benedict: At least one of us is.

Which of the two, if either, should be convicted?

Which of the two, if either, should be acquitted?

9 – A logician visiting this island at this time committed an act which was legally a capital offense there (he flirted with the king's wife). He was brought to trial and the king was the presiding judge. Now this king, though despotic in his own way, was nevertheless fair enough to give the accused some chance of going free. First he made a statement that was obviously true, and so the logician knew that he was either a healthy knight or a sick knave. Then the king pointed to four masked men and said to the logician: "One of those four may be my witch doctor. Each of the four will make a statement. If you are clever enough to deduce which one is the witch doctor or to prove that none of them are, then you may go free. But if you fail, you die!

The four made the following statements:

Awk: I am neither a healthy knight nor the witch doctor.

Barab: I am neither a sick knave nor the witch doctor.

Caleb: I am either a healthy knight or a sick knave or the witch doctor.

Dworg: I am neither a healthy knight nor a sick knave nor the witch doctor.

Fortunately, the logician was able to solve the problem, and so got off with his life. What is the solution? Are any of them the witch doctor? If not, why not? If so, which one?

SOLUTIONS

1 – A statement that works is: "I am a healthy knave". No truthful person would say that, hence the statement is false, and so the speaker must be a sick knight.

2 – A statement that works is: "I am not a healthy knight". A healthy knight wouldn't make such a false statement, a sick knight couldn't make such a true statement, and a healthy knave couldn't make such a true statement. Only a sick knave could make such a true statement.

3 – This is a bit more difficult: To say that an inhabitant is either sick or a knight (and possibly both) is tantamount to saying that he is not a healthy knave. So the problem can be restated: What statement could be made only by an inhabitant who is not a healthy knave? Well, a statement that works is: "I am not a sick knight". A sick knight could falsely say that, or a healthy knight could truthfully say that, or a sick knave could truthfully say that, but a healthy knave could not make that true statement.

4 – The solution is so simple that it is apt to be overlooked! To say that the native is either sick or a knight but not both, is to say nothing more nor less than that he is either a sick knave or a healthy knight–in other words that he is a truth-teller! And so a statement like "Two plus two is four" would do the trick.

5 – The simplest solution I can think of is: "I am sick". A sick knight couldn't make such a true statement; a healthy knight couldn't make such a false statement but a sick knave could truly say that, and a healthy knave could falsely say that. In summary, no knight could say that, but any knave–sick or healthy–could say that.

6 – A statement that works is: "I am either a healthy knight or a sick knave". This is tantamount to saying: "I am truthful", which a truthful inhabitant and a liar can both say.

7 – Aaron's statements are either both true or both false. If both true, then Bowaine is both a knight and sick, hence a sick knight, hence Bowaine is lying. If Aaron's statements are both false, then Bowaine is both a knave and healthy, so again Bowaine is lying. This proves that Bowaine's statements are both false, hence Aaron is both a knight and sick, hence Aaron also lied. Thus, all four statements are false, which means that Bowaine is a healthy knave and Aaron is a sick knight.

8 – If Archie were truthful, he would never make the false statement that he and Benedict were both sick knights, so Archie certainly lied. Thus the two are not both sick knights.

If Benedict's statement is false, then it is not the case that at least one of them is a healthy knave, but then both of them lied, hence both of them must be sick knights, which we have seen is not the case. Therefore, Benedict spoke the truth, hence at least one of them really is a healthy knave, but it can't be Benedict, who is truthful, so it must be Archie. Thus Benedict is truthful, hence not a sick knight, and Archie is a healthy knave, hence also not a sick knight, and so both are innocent and both should be acquitted.

9 – Awk is certainly not a healthy knight. He is either a sick knave and not the witch doctor, or a liar and the witch doctor. It cannot be determined from his statement just what he is.

From Barab's statement, it cannot be determined whether or not he is the witch doctor. (He could be a healthy knight and not the witch doctor, or a liar and the witch doctor.)

As to Caleb, his statement reveals practically nothing: He could be truthful (either a healthy knight or a sick knave), and if so, he might be the witch doctor, or again he might not be. He could also be lying, in which case he is not the witch doctor.

Dworg's statement, however, is definitive. He is certainly lying, since neither a healthy knight nor a sick knave could make such a statement. Thus Dworg is neither a healthy knight nor a sick knave, hence if he were also not the witch doctor, his statement would be true, which it isn't. Therefore, Dworg must be the witch doctor.

Chapter 11

Human or Android?

On a certain planet, half the inhabitants are human and the other half are androids who *look* like humans, but who have been artificially created in laboratories. Half of the humans are always truthful and half always lie. Same with the androids: half always tell the truth and half always lie.

A reporter from Earth named *Robert*, who was very good at logical reasoning, was once sent to this strange planet to make some investigations, and here are some of his encounters.

1 – One day he met an inhabitant who made a statement, from which Robert could deduce that the inhabitant must be a truthful human. What statement could accomplish this?

2 – On another occasion, Robert met an inhabitant who made a statement that conclusively proved that he was a lying android. What statement would work?

3 – Another time, an inhabitant made a statement from which the reporter could deduce that he was in the presence of an android, but there was no way of telling whether the statement was true or false. What statement could that have been?

4 – In a more puzzling incident, an inhabitant made a statement from which it follows that if the statement were true, then he must be an android, but if the statement were false, then he could be either a human or an android, and there is no way to tell which. What statement would work?

5 – On another occasion, Robert met an inhabitant who made a statement from which Robert could deduce that the speaker was either a truthful android or a lying human, but there was no way to tell which. What statement would work?

6 – On another occasion, Robert came across two inhabitants named *Auk* and *Bog*. They made the following statements:

Auk: We are not both androids.

Bog: We are not both truthful.

What are Auk and Bog?

7 – **Fingering an Android** — On another occasion, Robert came across three inhabitants named Cog, Dag, and Eg, one and only one of whom was human. They made the following statements:

Cog: Eg is human.

Dag: I am an android.

Eg: At most, one of us is truthful.

From these statements it is not possible to tell which one is human, but it is possible to determine one who is definitely an android. Which one? Also, was the human lying or telling the truth?

8 – On another occasion, Robert came across three natives named Hal, Jal, and Klak. Only one was an android. They made the following statements:

Hal: Jal is not a lying android.

Jal: If I am an android, then Hal is a liar.

Klak: The android is a liar.

Which one is the android?

9 – [A Metapuzzle] – On another occasion, Robert met two inhabitants named Zak and Yek, who made the following statements:

Yek: Zak is a truthful human.

Zak: Yek is a lying android.

From this it is not possible to tell what Yek and Zak both are. However, Robert later found out whether Zak told the truth or lied, and Robert then knew what each one was. What is each?

SOLUTIONS

1 – One statement that would work is: "I am not a truthful android". If the statement were false, then he would be a truthful android (contrary to what he said), hence he would be truthful, which is not possible if the statement is false. Therefore, he spoke truly, so he really is not a truthful android, and so he is a truthful human.

2 – A statement that works is: "I am a lying human". Obviously, the speaker was not truthful, so he is not really a lying human, but since he lied, he is a lying android.

3 – A statement that works is: "I am either a truthful android or a lying human".

Suppose the statement is true. Then he really is either a truthful android or a lying human, but he can't be a lying human since he is truthful and so he is then a truthful android.

On the other hand, suppose the statement is false. Then, he is neither a truthful android or a lying human. Since he is not a lying human, but he is lying, then he must be a lying android.

Thus, regardless of whether he told the truth or lied, he is an android. Whether or not he is truthful cannot be determined.

4 – The statement: "I am an android" won't work, because if true, then he is indeed an android, but if false, then he is definitely *not* an android.

However, the statement: "I am a truthful android" does work. If he is truthful, then, of course, he is a truthful android, but if the statement is false, he could be either a lying android or a lying human, and there is no way to tell which.

5 – This is really so simple that it is apt to be overlooked! All he need say is "I am an android". If he is truthful, then he is a truthful android, and if he is lying, then he is a lying human, and there is no way to tell which.

6 – If Bog were lying, then it would be true that not both are truthful, which is a contradiction. Hence Bog spoke the truth, and so it really is the case that they are not both truthful, which means that Auk lied, and so, contrary to what he said, both are androids. Thus, Auk is a lying android and Bog is a truthful android.

7 – **Case 1 — Eg told the truth**. Then it really is the case that at most one of the three is truthful, hence Cog and Dag both lied. Since Dag lied, then he is really human. So, in this case, Dag is a lying human.

Case 2 — Eg lied. Then it is not the case that at most one is truthful, hence at least two are truthful, and since Eg lied, it must be that Cog and Dag are both truthful. Since Cog told the truth, then Eg is human. And so, in this case, Eg is a lying human.

This proves that either Dag is human (Case 1) or Eg is human (Case 2), and so Cog is definitely an android. Also in both Case 1 and Case 2, the human lied.

In summary, Cog is definitely an android and the human definitely lied.

8 – **Step 1 — If Klak told the truth, then the android really is a liar, hence is not Klak. On the other hand, if Klak lied, then the android is not really a liar, hence again cannot be Klak. This proves that Klak is not the android.

Step 2 — Next we show that if Jal is an android, we get a contradiction. Well, suppose that Jal is an android.

Case 1 — Jal told the truth. Then it really is the case that if Jal is an android, then Hal lied, and since Jal *is* an android (by assumption) then Hal lied, which means that Jal *is* a lying android, contrary to the case under consideration, in which Jal is truthful.

Case 2 — Jal lied. Then Jal is a lying android, hence Hal's statement was false, hence Hal is a liar. Then it is *true* that *if* Jal is an android (which he is) *then* Hal is a liar, which means that Jal's statement was true, contrary to our assumption that he lied. Thus, this case is also out.

Thus, Jal is not the android, hence the android is Klak.

9 – If Yek is truthful, then Zak is truthful (as Yek said), hence Yek is a liar (as Zak said), which is a contradiction. Thus, Yek definitely lied. Now there are two cases to consider.

Case 1 – Zak told the truth. Then Yek is a lying android. Since Yek lied, then Zak is not a truthful human, but since he is truthful, he must be a truthful android. So in this case, both are androids.

Case 2 – Zak lied. Then Yek is not a lying android, but since Yek lied, he is a lying human. However, Zak could be either human or android (in both cases Yek's statement is false) and there is no way to tell which.

Now, if Robert had later found out that Zak lied, he couldn't have known whether Zak was human or android, but we are given that he *did* know what both were, hence he must have found out that Zak told the truth, and hence then knew that both were androids.

Chapter 12

Variable Lying and Paradox

Before leaving the subject of lying and truth-telling (which we will return to in a more formal manner in Book II), there are two more topics to be considered.

I – Variable Liars[1]

On the Island of Variable Liars, on each day, an inhabitant either lies the entire day, or tells the truth the entire day, but can change from one day to the next.

1 – **Boris and Michael** — Inhabitants Boris and Michael are two brothers who look so much alike that no one can visually tell them apart. However, Boris tells the truth only on Mondays, whereas Michael is truthful only on Tuesdays.

When Abercrombie came to this island, he one day came across these two brothers who made the following statements:

First One: I am Boris.

Second One: I am Michael.

Is it possible to tell from this which one was Boris? Is it possible to tell the day of the week?

2 – On another day, Abercrombie met the two brothers who made the following statements:

First One: I am Boris.

Second One: If that is true, then I am Michael.

Can it be determine who is who? Can the day of the week be determined?

---

[1]This topic will come up again in Book II.

3 – **A Metapuzzle** — Abercrombie knew that one of the brothers was an engineer, but he didn't know which one. He also knew the lying habits of both brothers.

One day, he met the two brothers and asked them which one was the engineer. They gave the following answers:

First One: Michael is the engineer.

Second One: I am Michael.

From this, Abercrombie, who was good at logic, couldn't determine whether it was Boris or Michael who was the engineer (although he, of course, knew the day of the week).

From what I have told you, you cannot tell whether the engineer was Boris or Michael, but you have enough information to determine whether it was the first one or the second one who was the engineer. Which one was it?

4 – **Arlene and Teresa** — Arlene and Teresa are twin sisters and are visually indistinguishable. Each of them is always dressed in either red or green. Arlene tells the truth when wearing green and lies when wearing red. Teresa is the opposite. She tells the truth when wearing red and lies when wearing green.

One day, a color-blind man named Alfred met one of the sisters and wanted to know what color dress she was wearing. What single yes/no question would determine this?

Suppose, instead, Alfred wanted to know whether he was addressing Arlene or Teresa. What yes/no question should he ask?

5 – **A Metapuzzle** — On another occasion, Alfred met one of the sisters and asked: "Are you Arlene who is now wearing red?" She answered (*yes* or *no*) and Alfred, who was a good logician, knew who she was. Who was she, and what color was she wearing?

6 – Boris was in love with Arlene and Michael was in love with Teresa. One day, one of the two brothers met one of the two sisters. She said: "Today is Tuesday". He said: "You are now wearing red". It is not recorded on what day of the week that this occurred, but it was either on a Monday or on a Tuesday.

Can it be determined whether the man was speaking to the lady he loved?

II – Truth, Falsehood and Paradox

7 – **What Goes Wrong?** — I often play the following trick on people, particularly logicians, who are especially apt to fall for it.

I show them two envelopes marked 1 and 2 and explain that one of them contains a dollar bill and the other does not. On the faces of the envelopes are written the following sentences:

1	2
The sentences on both envelopes are false.	The dollar bill is in the other envelope.

I tell the person that if he or she can determine which envelope contains the dollar bill, and can prove that it does before opening the envelope, then he or she can keep the dollar. So far, every one on whom I have tried this reasons as follows: "If the sentence on 1 is true, we obviously have a contradiction, hence it must be false. Therefore, it is not the case that both sentences are false, and since the first is false, the second must be true, and so the dollar bill is therefore in Envelope 1". The person then opens 1 and is surprised to see that the dollar bill is not there; it is then seen to be in Envelope 2.

What is wrong with the reasoning? [The solution to this should be read *before* the next problem.]

8 – There is a yes/no question I could ask you such that if you answer *yes*, you are right, and if you answer *no*, you are also right! What question would work?

9 – There is another yes/no question I could ask you such that if you answer *yes*, your answer would be neither true nor false, but paradoxical, and likewise if you answer *no*. It is impossible for you to avoid answering paradoxically! What question would work?

SOLUTIONS

1 – If either statement is true, so is the other, hence they are either both true or both false. They cannot both be true, since the two brothers are never truthful on the same day, hence they are both false. Thus, the first one is really Michael and the second one is Boris. The day of the week cannot be determined. All that can be said is that it was neither Monday nor Tuesday.

2 – What the second one said is obviously true. Therefore, the first one lied (since the two brothers are never both truthful on the same day), hence he is Michael. Thus, the second one is Boris, and hence the day must be Monday, the only day that Boris is truthful.

3 – If the day had been other than Monday or Tuesday, Abercrombie would have known that both brothers were lying, hence that the first one was lying, and therefore that Boris was the engineer.

Had the day been Monday, Abercrombie would have deduced that Michael was the engineer as follows: If the first one was lying, he would have to be Michael (since Boris is truthful on Monday), hence the second one would be Boris, who made the false claim to be Michael, contrary to the fact that Boris is truthful on

Monday. Therefore, the first one couldn't be lying; he must be truthful, and so Michael is the engineer.

Thus, if the day were any day other than Tuesday, Abercrombie would have known whether the engineer was Boris or Michael, but he didn't know, and therefore the day must be Tuesday.

Now, you and I know that the day was Tuesday. The first possibility is that the first was truthful. In this case, he must be Michael (because Boris lies on Tuesdays) and also Michael must really be the engineer (as he truthfully said). So in this case, the first one is the engineer. The second possibility is that the first one lied. In this case, he must be Boris (since Michael is truthful on Tuesdays), and also Boris must be the engineer (contrary to Boris' false claim), so again the first one is the engineer.

In summary, the day was Tuesday, the first one was the engineer (who could be either Michael telling the truth, or Boris, lying).

4 – This is old hat: To find out if she is wearing green, he could ask her if she is Teresa. To find out if she is Teresa, he could ask her if she is wearing green.

5 – If she had answered *no*, she could be either Arlene in green, Arlene in red or Teresa in green, and Alfred could have no way of knowing which. Therefore, she must have answered *yes*, and Alfred then knew that there was no possibility other than she was Teresa in red.

6 – **Case 1 – The man was Boris** — If the day was Monday, then he told the truth, hence she was really dressed in red, but she lied in saying that it was Tuesday, hence she is then Arlene (who lies, when in red). On the other hand, if the day was Tuesday, then he lied, hence she was really dressed in green, but she told the truth in saying it was Tuesday, hence again she is Arlene. Therefore, if the man was Boris, he was speaking to the lady he loved.

Case 2 – The man was Michael — If the day was Monday, then he lied, hence she was really wearing green, and she also lied in saying it was Tuesday, hence she must then be Teresa. Now, suppose it was Tuesday. Then both told the truth. (Michael tells the truth on Tuesdays), hence she really was wearing red, and so again she is Teresa. Therefore, if the man was Michael, the lady was Teresa, and so he was speaking to the lady he loved.

This proves that regardless of whether the man was Baris or Michael, he was speaking to the lady he loved.

7 – What is wrong is the assumption that every sentence must be either true or false! Some sentences can be neither true nor false without involving a contradiction. Such sentences are called *paradoxical*. A Typical example would be a single sentence written on a page which says: "The sentence written on this page is false". If it were true, what it says would be the case, which would mean that it is false, which is a contradiction. On the other hand, if it were false, then what it says is

not the case, which would mean that it is not false, which is again a contradiction. Thus, the sentence cannot be either true or false without contradiction and so it is paradoxical[2].

Coming back to the two envelopes, the second sentence was not paradoxical, but simply false (since the bill was not in Envelope 1). As to the first sentence, it was not exactly paradoxical, but nevertheless, neither true nor false, since its truth would involve a logical contradiction, whereas its falsity wouldn't involve a *logical* contradiction, but only the *factually* false proposition that the bill was in Envelope 1. At any rate, the sentence on 1 was neither true nor false.

8 – Will you answer *yes* to this question?

9 – Is *no* the correct answer to this question?

---

[2]A whole bunch of interesting paradoxes can be found in the last chapter of my previous book, *The Riddle of Scheherazade.*

BOOK II
The Magic Garden

Chapter 13

George's Garden

We now turn to a remarkable problem whose solution will take us through an amazing labyrinth of smaller problems leading us into the very heart of a theory which has proved extremely important.

A – We Enter the Garden

A Grand Problem — A little boy named George B. had a garden of magic flowers. Each flower could change color from day to day, but it assumed only two colors–red or blue (just like litmus paper). On each day the flower would be either red the entire day or blue the entire day. We are given that the following condition holds.

Condition B — For any flowers A and B–whether the same or different–there is a flower C which is red on those and only those days on which A and B are both blue. (Thus on any day on which at least one of the flowers A or B is red, C is blue, but an any day on which A and B are both blue, the flower C is red).

Two flowers are said to be *similar*, or to be of the *same type* if they are of the same color on all days. Well, one day George decided to prune his garden, and so removed many of the flowers. After the garden was pruned, Condition B still held, but in addition, no two distinct remaining flowers were of the same type. Thus, given any two *distinct* flowers, there was at least one day on which one of them was red and the other was blue.

Now comes a "grand" problem! The number of flowers in the garden after pruning was somewhere between two hundred and five hundred. How many flowers were there?

It may well seem incredible to many readers that this problem has a solution, but it will be seen that it really has! I call this a *grand* problem because we need many subsidiary problems to solve it, some of which will be given in this chapter. The final solution won't emerge until the end of Chapter 15 (which could be read directly following this chapter). In this chapter, we will derive a host of useful consequences of Condition B that will be basic to several chapters that follow. None of the results of this chapter depend on the fact that George pruned his garden, and so we now go back to the days *before* the pruning (though all results also hold for the garden after the pruning).

A Preliminary Question — Does it follow from Condition B that for any flowers A and B there is a flower that is blue on those and only those days on which A and B are both red? (The reader might try to answer this now, but the answer will emerge in the course of the next few problems).

Problems 1-10 — Amazingly enough, the laws $B_1 - B_{10}$, below all follow as consequences of just Condition B, above. How do they follow? [Solutions follow B_{10}].

B_1: For every flower A there is a flower that is blue on those days when A is red, and red on those days that A is blue (and thus is of different color to A on all days). [Hint: Condition B holds even when A and B happen to be the same flower].

B_2: For any flowers A and B there is a flower which is blue on those and only those days on which A and B are both blue.

B_3: For any flowers A and B there is a flower that is blue on those and only those days on which at least one of the flowers A or B is blue (and thus is red on those and only those days on which A and B are both red).

B_4: For any flowers A and B there is a flower that is blue on those and only those days on which either A is red or B is blue (or both). [Thus, the flower is red on only those days on which A is blue and B is red].

B_5: For any flowers A and B there is a flower that is blue on those and only those days on which A and B are of the same color.

B_6: For any flowers A and B there is a flower that is blue on those and only those days on which A and B are of different colors (one is red and the other is blue).

B_7: For any flowers A and B there is a flower that is blue on those and only those days on which A is blue and B is red.

B_8: For any flowers A and B there is a flower that is blue on those and only those days on which A and B are both red.

B_9: At least one flower is blue on all days.

B_{10}: At least one flower is red on all days.

1 – We are given that for any flowers A and B, *whether the same or different*, there is a flower C that is red on those and only those days on which A and B are both blue. Among all such possible flowers C, we shall single out one of them and dub it $A \mid B$ (read "A stroke B"). Thus, $A \mid B$ is red on those and only those days on which A is blue and B is blue.

Now, what about the flower $A \mid A$? How does that behave? Well, $A \mid A$ is red on exactly those days on which A is blue and A is blue, but to say "A is blue and A is blue" is but a redundant way of simply saying that A is blue. Thus $A \mid A$ is red on those and only those days on which A is blue. Thus $A \mid A$ is a flower that solves the problem. We shall henceforth abbreviate $A \mid A$ by the single symbol $\overline{A}$ (read "A bar"). We thus rewrite law B_1 in the following symbolic form:

B_1: $\overline{A}$ is blue on those days on which A is red and red on those days on which A is blue. (Thus $\overline{A}$ is always of a different color than A).

2 – Given flowers A and B, let C be the flower $A \mid B$. How does $\overline{C}$ behave? Well, since C is red on just those days that A and B are both blue, then $\overline{C}$ is blue on just those days that A and B are both blue. Thus $\overline{C}$–which is $\overline{A \mid B}$ is the solution to our problem. We henceforth abbreviate $\overline{A \mid B}$ by $A \cap B$ (read "A cap B") and rewrite law B_2 symbolically thus:

B_2: $A \cap B$ is blue on those and only those days on which A and B are both blue.

3 – We let $A \cup B$ (read "A union B") be the flower $\overline{A} \mid \overline{B}$. It is red on those and only those days on which $\overline{A}$ and $\overline{B}$ are both blue, and thus is red an those and only those days on which A and B are both red. Therefore, $A \cup B$ is blue on just those days on which A and B are not both red–that is, on just those days when at least one of the flowers A or B is blue.

We now rewrite law B_3 thus:

B_3: $A \cup B$ is blue on those and only those days on which either A or B is blue (or both).

Notes – The *cap* operation $\cap$ is closely related to the notion of *and* (conjunction) in logic, since $A \cap B$ is blue just when A is blue *and* B is blue, whereas the *union* operation $\cup$ is related to the notion of *or* (disjunction) in logic, since $A \cup B$ is blue just when A is blue *or* B is blue (or both). The *bar* operation is related to the notion of *not* (negation) in logic, since $\overline{A}$ is blue just when A is *not* blue.

These operations are also related to the fundamental operations in the field known as *Boolean algebra*, that we will study later on.

B_4: We let $A \supset B$ (read "A imp B") be the flower $\overline{A} \cup B$. It is clearly blue on just those days on which either A is red, or B is blue (or both).

Thus, $A \supset B$ is blue on all days that A is red, and also on all days that B is blue (and also, of course, on all days when A is red *and* B is blue). The only days that $A \supset B$ is red are those on which A is blue and B is red (and is red on all such days).

The operation *imp* is closely related to *implication* in logic. To say that $A \supset B$ is blue on a given day is to say that *if* A is blue on that day, so is B (because if B were red on that day, then A would be blue and B red, hence $A \supset B$ would be red). Also, if $A \supset B$ happens to be blue on all days, then B is blue on all days that A is blue.

We rewrite law B_4 thus:

B_4: $A \supset B$ is blue on just those days when either A is red or B is blue (or both).

5 – We let $A \equiv B$ be the flower $(A \cap B) \cup (\overline{A} \cap \overline{B})$. It is blue on just those days on which either $A \cap B$ is blue (which is when A and B are both blue) or when $(\overline{A} \cap \overline{B})$ is blue (which is when A and B are both red). Thus we have:

B_5: $A \equiv B$ is blue on just those days when A and B are of the same color.

Note: We could alternately have taken $(A \supset B) \cap (B \supset A)$ for $A \equiv B$.

The operation of $\equiv$ is related to the notion of *equivalence* in logic: Two propositions are called *equivalent* if each implies the other–which means that they are either both true or both false, or stated otherwise, either one is true *if and only if* the other is true. And so, to say that $A \equiv B$ is blue on a given day is tantamount to saying that on that day, A is blue *if and only if* B is blue.

6 – We let $A \not\equiv B$ be the flower $\overline{A \equiv B}$. Thus $A \not\equiv B$ is always of different color from $A \equiv B$, and so we have:

B_6: $A \not\equiv B$ is blue on just those days when A and B are of different color (one red, and the other, blue).

7 – Obviously, the flower $A \cap \overline{B}$ does the job–it is blue just when A is blue and B is red (since it is blue just when A is blue and $\overline{B}$ is blue). We henceforth let A-B be the flower $A \cap \overline{B}$ and rewrite B_6 thus:

B_7: A-B is blue on those and only those days on which A is blue and B is red.

We might note that A-B is always of a different color from $A \supset B$. Thus $A \supset B$ is similar to $\overline{A - B}$ (They are always of the *same* color).

8 – We let $A \downarrow B$ be the flower $\overline{A} \cap \overline{B}$ and we obviously have:

B_8: $A \downarrow B$ is blue on those and only those days on which A and B are both red (since these are the days on which $\overline{A}$ and $\overline{B}$ are both blue).

9 – For any flower A, the flower $A \cup \overline{A}$ is always blue (because on any day, one of the flowers A or $\overline{A}$ is blue and the other is red, and so in either case, $A \cup \overline{A}$ is blue). And so we pick one flower A at random (it makes no difference which one) and take b to be the flower $A \cup \overline{A}$, and we thus have:

B_9: b is always blue.

10 – Obviously, $\overline{b}$ is always red, and so we take r to be $\overline{b}$, and we have:

B_{10}: r is always red.

Note: We could alternatively have taken r to be $A \cap \overline{A}$, where A can be any flower.

B – Some Boolean Relationships

Now that the reader is familiar with the meanings of $\overline{A}, A \cap B, A \cup B$, $A \supset B, A \equiv B, A \not\equiv B, A - B, A \downarrow B, A \mid B$, and b (a flower always blue) and r (a flower always red), we can consider some relations between flowers.

We recall that we are calling two flowers A and B *similar* if they are the same color on all days. Now, suppose I asked you a question such as whether $\overline{A \cap B}$ is necessarily similar to $\overline{A} \cup \overline{B}$, how would you go about finding out? Here is one *systematic* way of answering questions of this type: On any day, there are two possibilities for the color of A–blue or red. With each of these possibilities for A, there are the same two possibilities for B, and so there are four possibilities altogether: (1) A blue and B blue; (2) A blue and B red; (3) A red and B blue; (4) A red and B red. Thus, on any day, one and only one of these four possibilities holds. In each case, you can tell the color of $\overline{A \cap B}$ and of $\overline{A} \cup \overline{B}$ and see if they are the same. If they are the same in all four cases, then you know that the two flowers in question are similar.

Let us try this: In Case(1) A and B are both blue, hence $A \cap B$ is blue, hence $\overline{A \cap B}$ is red. On the other hand, since A and B are both blue, then $\overline{A}$ and $\overline{B}$ are both red, hence $\overline{A} \cup \overline{B}$ is red. And so on a day when A and B are both blue, $\overline{A \cap B}$ and $\overline{A} \cup \overline{B}$ are both red, hence they are the same color on such a day.

Next, let's consider a day when A is blue and B red (Case 2). Then $A \cap B$ must be red, hence $\overline{A \cap B}$ is then blue. On the other hand, looking at $A \cup B$, we see that on such a day, $\overline{A}$ is red and $\overline{B}$ is blue (since A is blue and B is red), hence $\overline{A} \cup \overline{B}$ is then blue. And so in Case(2), $\overline{A \cap B}$ and $\overline{A} \cup \overline{B}$ are again of the same color–both blue.

The reader can now verify that in Case(3) both flowers are blue, and likewise with Case(4). Thus the flowers are the same color in all four possible cases, and so they are similar.

This type of argument might be termed *analysis by cases*. Its virtue is that it is completely systematic and always works. However, a more creative solution of the above problem is this: On any day that $\overline{A \cap B}$ is blue, $A \cap B$ must be red, hence either A is then red or B is then red (or both), hence either $\overline{A}$ is then blue or

$\overline{B}$ is then blue, bence in either case $\overline{A} \cup \overline{B}$ is then blue. This proves that whenever $\overline{A \cap B}$ is blue, so is $\overline{A} \cup \overline{B}$. Going in the other direction on any day that $\overline{A} \cup \overline{B}$ is blue, either $\overline{A}$ is blue or $\overline{B}$ is blue, hence A is then red or B is then red, hence $A \cap B$ is then red, hence $\overline{A \cap B}$ is then blue, and so whenever $\overline{A} \cup \overline{B}$ is blue, so is $\overline{A \cap B}$. Thus, whenever either of the two flowers $\overline{A \cap B}, \overline{A} \cup \overline{B}$, is blue, so is the other, and so they are similar.

Sometimes a little common sense works best. For example, to show that $A \cap B$ is similar to $B \cap A$, rather than go to the labor of a case analysis, we can immediately see that $A \cap B$ is blue when and only when A and B are both blue, but to say that A and B are both blue is the same as saying that B and A are both blue, and this occurs when and only when $B \cap A$ is blue. So, of course, $A \cap B$ is similar to $B \cap A$. It is equally obvious that $A \cup B$ is similar to $B \cup A$. But in cases which are not immediately obvious, we can always rely on a case analysis.

We strongly recommend that the reader works through the following exercises:

Exercise 1 — Verify by any reasonable means the following similarities:

(1) $A \supset B$ is similar to $\overline{B} \supset \overline{A}$

(2) A is similar to $\overline{\overline{A}}$

(3) $\overline{A \supset B}$ is similar to $A \cap \overline{B}$

(4) $A \cup \overline{A}$ is similar to b

(5) $A \cap \overline{A}$ is similar to r

(6) $A \cap (A \cup B)$ is similar to A

(7) $A \cup (A \cap B)$ is similar to A

(8) $A \cap b$ is similar to $A \cup r$

(9) $A \supset B$ is similar to $A \equiv (A \cap B)$

(10) $A \cap B$ is similar to $A \equiv (A \supset B)$

(11) $\overline{A \cup B}$ is similar to $B \equiv (A - B)$

(12) A-B is similar to $\overline{B} \equiv (A \cup B)$

Exercise 2 — Prove that the following flowers must be blue on all days.

(1) $A \supset (B \supset A)$

(2) $(A \cap B) \supset A$

(3) $A \supset (A \cup B)$

(4) $r \supset r$

(5) $(\overline{A \cup B}) \equiv \overline{A} \cap \overline{B}$

Exercise 3 — Let us call two flowers A and B *independent* if there is one day on which they are both blue; one day on which they are both red; one day on which A is blue and B is red; and one day on which A is red and B is blue.

Now, suppose A and B are flowers such that A is blue on Mondays and Tuesdays, but red on Wednesdays and Thursdays, whereas B is blue on Mondays and Wednesdays, but red on Tuesdays and Thursdays. Then, clearly A and B are independent. In each of the pairs of flowers below, the pair is definitely *not* similar. In each case, determine on which of the first four days of the week, the two flowers are of different colors.

Pair (1) – A and B

Pair (2) – $A \supset B$ and $B \supset A$

Pair (3) – $A \cap B$ and $A \supset B$

Pair (4) – $A \supset B$ and A

Pair (5) – $A \supset B$ and B

Pair (6) – $\overline{A \cap B}$ and $A \cup B$

[Answers to Exercise 3 are given at the end of the next chapter].

Chapter 14

Some Neighboring Gardens

We shall now explore some other flower gardens, closely related to George's garden, and thus see some important inter-relationships between the laws B, B_1 - B_{10} of the last chapter.

A – Some Other Boolean Gardens

We shall define a flower garden to be a *Boolean garden* iff[1] it satisfies condition B of Chapter 13, which we have seen also implies the conditions B_1 - B_{10}, which it will be convenient to now review.

B: $A \mid B$ is blue on those and only those days when at least one of A or B is red (or what is the same thing, $A \mid B$ is red on those and only those days on which A and B are both blue.

B_1: $\overline{A}$ is blue on those days when A is red, and red on those days on which A is blue.

B_2: $A \cap B$ is blue on just those days when A and B are both blue.

B_3: $A \cup B$ is blue on just those days when at least one of A or B is blue (or both). [Thus, $A \cup B$ is red when and only when A and B are both red].

B_4: $A \supset B$ is blue on just those days on which A is red or B is blue (or both).

B_5: $A \equiv B$ is blue on just those days on which A and B are of the same color.

B_6: $A \not\equiv B$ is blue on just those days when A and B are of different colors.

B_7: $A - B$ is blue on just those days when A is blue and B is red.

---

[1] We write "iff" to mean "if and only if". This is a useful and fairly common abbreviation in mathematics.

B_8: $A \downarrow B$ is blue on those and only those days on which A and B are both red.

B_9: b is blue on all days.

B_{10}: r is red on all days.

Thus, a Boolean garden satisfies conditions B_1-B_{10}, as well as B. Now, the point is that there are several other ways of obtaining a Boolean garden than by *starting* with condition B. That is, there are several different groups of some of the laws B_1-B_{10} which in turn imply condition B (and hence all the other laws as well). All this related to the mathematically vital subjects of Propositional Logic and Boolean Algebra, which we will look at in later chapters.

Now, let us look at several other ways of obtaining a Boolean garden than starting with Condition B. As with George's garden, in all the gardens to be described in this chapter, each flower is either blue the entire day, or red the entire day, though it might change its color from day to day. In this chapter we needn't assume that the garden is pruned.

1 – The Garden of Charles — George's friend Charles has a garden in which to every flower A there corresponds a flower $\overline{A}$ which is always of different color than A (law B_1) and to any flowers A and B there corresponds a flower $A \cap B$ which is blue on those and only those days on which both of the flowers A and B are blue (law B_2). Prove that Condition B of George's garden must hold, and hence that Charles' garden is also a Boolean garden.

2 – David's Garden — George has another friend David who has a garden satisfying laws B_1 and B_3. Prove that this garden is also a Boolean garden.

3 – Irving's Garden — In Irving's garden, laws B_1 and B_4 are known to hold. Prove that this garden is a Boolean garden.

4 – The Garden of Irving's Brother — Irving's brother has a garden in which law B_4 holds and in addition there is one flower r that is red all the time (law B_{10}). Is this garden necessarily a Boolean garden?

5 – Jerry's Garden — In Jerry's garden, all that is given is that the law B_8 holds. Is this garden necessarily a Boolean garden?

6 – Edward's Garden — It is known that laws B_2, B_3, B_4 and B_5 hold for Edward's garden. Is this garden necessarily a Boolean garden?

7 – Michael's Garden — In this garden, laws B_1 and B_7 hold. Is Michael's garden necessarily a Boolean garden?

8 – The Garden of Michael's Brother — In this garden, laws B_7 and B_9 are known to hold. Is this garden necessarily a Boolean garden?

B – Some Partial Boolean Gardens

We have so far studied Boolean gardens–gardens in which all the laws B_1-B_{10} hold. Now, we might have a garden in which some but not necessarily *all* of these ten laws hold. Such gardens we call *partially* Boolean, or *partial Boolean gardens*. The interesting thing about these gardens is that even though not all the laws follow from the given ones, some others might. The interrelationships can be quite fascinating.

Now let us consider some *partial* Boolean gardens.

9 – The Garden of Eve — George's friend Eve has a garden in which laws B_4 and B_5 are known to hold. It is impossible from just this information to prove that all the laws B_1 - B_{10} hold, but it is possible to derive law B_2. How? [The solution is quite tricky!]

10 – Evelyn's Garden — Eve's sister Evelyn has a garden in which B_2 and B_5 hold. Prove that law B_4 must also hold.

11 – Erika's Garden — Eve has another sister Erika, whose garden satisfies laws B_2 and B_4. Prove that it also satisfies law B_5.

Note: We see from the last three problems that laws B_2, B_4 and B_5 form an interesting trio: If any two of them hold for a given garden, the third must hold as well.

12 – Another Trio — Another interesting trio consists of the laws B_3, B_7 and B_6: Any garden that, obeys any two of them must obey the remaining one as well. Prove this.

13 – Irma's Garden — All that is given about Irma's garden is that it obeys B_4. Prove that it also obeys B_3. [The solution is tricky!]

14 – Mary's Garden — All that is given about this garden is that it obeys law B_7. Prove that it obeys B_2.

15 – The Garden of Mary's Sister — This garden obeys B_7 and B_6. Prove that it obeys B_3.

16 – Lillian's Garden — Lillian's garden obeys laws B_3 and B_5. Prove that it obeys B_2. [This is a bit tricky!]

17 – Lillian's garden also obeys B_4. Prove this.

18 – The Garden of Lillian's Sister — This garden obeys laws B_2 and B_6. Prove that it obeys B_3.

SOLUTIONS

1 – For any flowers A and B, we have the flower $A \cap B$, which is blue on just those days in which A and B are both blue, and then we have the flower $\overline{A \cap B}$, which is red on just those days. And so we define $A \mid B$ to be the flower $\overline{A \cap B}$, and this "stroke" operation satisfies condition B.

2 – For David's garden, we take $A \mid B$ to be the flower $\overline{A} \cup \overline{B}$. It is blue when at least one of the flowers $\overline{A}, \overline{B}$, is blue, thus it is blue when at least one of A, B is red–hence it is red on just those days that A and B are both blue. Thus, condition B holds.

3 – For Irving's garden, take $A \mid B$ to be $A \supset \overline{B}$. It is red just when A and B are both blue, so condition B holds.

4 – Yes, it is: We first take $\overline{A}$ to be $A \supset r$ and so $\overline{A}$ is always of different color than A (as is easily verified). Thus, we have law B_1, and since we are given law B_4, the laws of David's garden hold, which we have seen to give condition B. Or, more directly, given laws B_4 and B_{10}, we define $A \mid B$ to be $A \supset (B \supset r)$, and condition B is then satisfied $(A \supset (B \supset r)$ is red on just those days on which A and B are both blue, as the reader can verify).

5 – For Jerry's garden, we first take $\overline{A}$ to be $A \downarrow A$ and $\overline{A}$ is then always of different color to A. We then take $A \cap B$ to be $\overline{A} \downarrow \overline{B}$. Then we take $A \mid B$ to be $\overline{A \cap B}$ (as in Problem 1) and we have Condition B $(A \mid B$ is red on just those days that A and B are both blue). Thus Jerry's garden is also a Boolean garden.

6 – Edward's garden *might* be a Boolean garden, for all we know, but from just laws B_2, B_3, B_4 and B_5, we cannot deduce that it is. For all we know, it might be that all of Edward's flowers are blue all the time, in which case laws B_2, B_3, B_4 and B_5 obviously hold (as well as law B_9), but B_{10} fails, which cannot happen in a Boolean garden.

7 – The flower $A - \overline{B}$ is blue just when A and B are both blue, hence we have law B_2. Also we have law B_1 given, and so Michael's garden satisfies the conditions of Charles' garden, which we already know to be a Boolean garden. [More directly, the flower $\overline{A - \overline{B}}$ is red on those and only those days on which A and B are both blue, and so we immediately have Condition B, taking $A \mid B$ to be $\overline{A - \overline{B}}$].

8 – The flower $b - A$ is blue on just those days when b is blue and A is red, but since b is always blue, these days are simply those on which A is red. Thus, given the dash operation, we can take $\overline{A}$ to be $b - A$, and so $\overline{A}$ is always of different color than A. We thus have law B_1, which together with B_7, insures that the garden is Boolean (as shown in the solution of the last problem).

9 – Given the operations $\supset$ and $\equiv$, we can take $A \cap B$ to be $A \equiv (A \supset B)$, which is blue on those and only those days on which A and B are both blue (as the reader can verify).

10 – We are now given the operations $\equiv$ and $\cap$, and so we take $A \supset B$ to be $A \equiv (A \cap B)$, which is blue just when either A is red or B is blue (or both).

11 – Given the operations $\supset$ and $\cap$, we obviously take $A \equiv B$ to be $(A \supset B) \cap (B \supset A)$.

12 – (1) Suppose we are given B_3 and B_7–thus we have the operations $\cup$ and $-$. Then take $A \not\equiv B$ to be $(A - B) \cup (B - A)$, which is blue just when A and B are of different color. We then have law B_6.

(2) Suppose we are given B_3 and B_6–thus the operations $\cup$ and $\not\equiv$. Then we take $A - B$ to be $B \not\equiv (A \cup B)$, which is blue just when A is blue and B is red. Thus, we have B_7.

(3) Now suppose we are given B_6 and B_7. Then we take $A \cup B$ to be $B \not\equiv (A - B)$, which is blue just when at least one of the flowers A or B is blue. Thus we then have B_3.

13 – We are given the operation $\supset$ and we are to derive the operation $\cup$. Well, take $A \cup B$ to be $(A \supset B) \supset B$, which can be seen to be blue on just those days when at least one of A or B is blue.

14 – The flower $A - (A - B)$ is blue on just those days when A and B are both blue, hence we can take $A \cap B$ to be $A - (A - B)$.

15 – Take $A \cup B$ to be $A \not\equiv (B - A)$.

16 – Take $A \cap B$ to be $(A \cup B) \equiv (A \equiv B)$.

17 – Take $A \supset B$ to be $(A \cup B) \equiv B$.

18 – Take $A \cup B$ to be $(A \cap B) \not\equiv (A \not\equiv B)$.

Answers to Exercise 3, Chapter 13

(1) Tuesday and Wednesday

(2) Tuesday and Wednesday

(3) Wednesday and Thursday

(4) Tuesday, Wednesday and Thursday

(5) Thursday

(6) Monday, Tuesday, Wednesday and Thursday

Chapter 15

The Grand Problem Solved!

This chapter will be devoted to working out the solution of the "grand" problem raised in Chapter 13.

The Grand Solution

We recall that we were given condition B of George's garden (for any flowers A and B there is a flower C that is red on those and only those days on which A and B are both blue). Such a garden we have termed a *Boolean* garden. We were then given that after George pruned his garden (so that for any two *distinct* remaining flowers, there was at least one day on which they were of different colors) and counted the remaining flowers, the number was somewhere between 200 and 500, and the problem is to determine how many flowers there were. The solution to this remarkable problem involves first solving many preliminary problems, to which we shall turn.

Of the ten laws B_1-B_{10}, which we have seen to follow from the single condition B, the following five are most relevant to this chapter.

B_1: $\overline{A}$ is always of different color from A.

B_2: $A \cap B$ is blue on those days when A and B are both blue, and red on all other days.

B_3: $A \cup B$ is blue on the days when at least one of the flowers A or B is blue, and is red on those and only those days when A and B are both red.

B_9: The flower b is blue on all days.

B_{10}: The flower r is red on all days.

Now we consider the situation as it is *after* George prunes his garden. We then have the following additional condition.

∪: [uniqueness condition] – For any flowers A and B, if A and B are of the same color on all days, then A and B are the same flowers. Stated otherwise, if flower A is *different* from flower B, then there is at least one day on which they are of different colors.

We shall call a Boolean garden a *reduced* Boolean garden if it obeys this uniqueness condition. We recall that we are calling two flowers *similar* if they are the same color on all days. Thus, a reduced Boolean garden is a Boolean garden in which similar flowers are identical.

It follows from this uniqueness condition of a reduced Boolean garden that b is the *only* flower that is blue on all days; r is the *only* flower that is red on all days. Also, for any flower A, the flower $\overline{A}$ is the *only* flower that is always different in color from A. Also, for any flowers A and B, the flower $A \cap B$ is the *only* flower that is blue on just those days when A and B are both blue, and $A \cup B$ is the *only* flower that is blue on those and only those days on which at least one of the flowers A or B is blue.

Now we shall turn to some fundamental definitions and auxiliary problems necessary for the solution of the "grand problem".

Domination — We will say that flower A *dominates* flower B if B is red on all days that A is red, or what is the same thing, A is blue on all days that B is blue. The only way that A can fail to dominate B is that there is at least one day on which B is blue and A is red.

Now let us consider the flower r which is red on all days. Then, obviously, every flower dominates r. Also r doesn't dominate any flower other than r itself (because if r dominates A, then A is red whenever r is red, which is all the time; hence A is always red; hence $A = r$ by the uniqueness condition ∪).

Next, we note that the flower b, which is always blue, dominates every flower A (because b is certainly blue whenever A is blue, being blue all the time). Also, every flower dominates itself. It is also obvious that if A dominates B and B dominates C, then A dominates C. Also, if A dominates B and B dominates A, then A and B must be the same flower (because each one is red when the other is red; hence they are always of the same color; hence are identical by the uniqueness condition). Another obvious fact is that for any flowers A and B, both A and B dominate $A \cap B$, and both A and B are dominated by $A \cup B$.

Let us record these simple facts.

Fact a – Every flower dominates r.

Fact b – r dominates no flower other than r itself.

Fact c – b dominates all flowers.

Fact d – Every flower dominates itself.

Fact e – If A dominates B and B dominates C, then A dominates C.

Fact f – If A and B dominate each other, then $A = B$.

Fact g – A and B both dominate $A \cap B$, and are both dominated by $A \cup B$.

No Domination Cycles — By Fact f, there cannot be two *distinct* flowers A and B, each of which dominates the other. Also, there cannot be three *distinct* flowers A, B and C such that A dominates B, B dominates C and C in turn dominates A, because if A dominates B and B dominates C, then A dominates C (Fact e), hence C cannot dominate A (Fact f). Likewise, there cannot be four *distinct* flowers A, B, C, D such that dominates B, which dominates C, which dominates D, which in turn dominates A, because A would then dominate D (by two applications of Fact e), and we would have A and D dominating each other, contrary to Fact f. And in general, we cannot have a finite sequence $A_1, A_2, \ldots, A_n$ of *distinct* flowers such that each one, other than the last, dominates the next, and the last one A_n, in turn dominating the first A_1. [Such a sequence, if it existed, would be called a *domination* cycle]. And so we have:

Fact h – There are no domination cycles.

Basic Flowers — Now comes a fundamental definition. By a *basic* flower we shall mean a flower X *other than* r which dominates no flower other than itself and r. Thus if X is basic, $X \neq r$, and for any flower Y, if X dominates Y, then either $Y = X$ or $Y = r$.

Now, suppose that X is unequal to r and that X is not basic. If r were the *only* flower other than X which X dominated, then X would be basic, which it isn't, and therefore X must dominate some flower Y other than X and r. And so, we have the following fact:

Fact(i) – If $X \neq r$ and X is not basic, then X dominates some flower Y other than X and r.

Now, for some problems whose solutions lead up to the solution of the grand problem.

Problem 1 — Consider the following three conditions.

(1) F dominates G.

(2) $G \cap F = G$

(3) $\overline{F} \cap G = r$

Does any one of these conditions imply any of the others?

Problem 2 — Is it possible for a flower other than r to be dominated by both F and $\overline{F}$?

Problem 3 — Is it possible for one basic flower to dominate another basic flower?

Problem 4 — Prove that if F fails to dominate X and X is basic, then $X \cap F$ is always red.

Problem 5 — Prove that for any *basic* flower X, and any flower F, either F dominates X or $\overline{F}$ dominates X (but not both).

Problem 6 — Is it possible for two *distinct* basic flowers to be blue on the same day?

Problem 7 — Suppose X is a basic flower. Are either of the following statements necessarily true?

  (a) There must be at least one day on which X is red and all the other basic flowers are blue.

  (b) There must be at least one day on which X is blue and all the other basic flowers are red.

Problem 8 — Show that every flower other than r dominates at least one basic flower. [Note: The proof depends on the fact that there are only *finitely* many flowers in George's garden. The result would not necessarily hold for an infinite Boolean garden].

Problem 9 — [Crucial!] – Prove that if F and G are *distinct* flowers, then one of them must dominate some basic flower which is not dominated by the other.

    For any flower F, we let F^* be the set of all *basic* flowers dominated by F. By Proposition 9 (stated following the solution of Problem 9), we at once have

Proposition A — If $F \neq G$, then $F^* \neq G^*$.

    Proposition A says that if F and G are distinct flowers, then the set of basic flowers dominated by F is different from the set of basic flowers dominated by G. Proposition A can be equivalently stated: If $F^* = G^*$, then $F = G$.

Basic Sets and Their Representatives — By a *basic set* we shall mean a set of basic flowers. For any basic set S and any flower F, we shall say that F *represents* S, or that F is a *representative* of S, if F dominates all the flowers in S and no other basic flowers–in other words, if $S = F^*$. A basic set S, if it has any representative at all, cannot have more than one representative, because if S were represented by F and by G, then $S = F^*$ and $S = G^*$, hence $F^* = G^*$, hence $F = G$ by Proposition A. We thus have:

Proposition B — A basic set cannot have more than one representative.

    Now comes a vital question: Does every basic set necessarily have a representative? We now turn to this.

Finite Unions – We know that for any two flowers F_1, and F_2, there is a flower that is blue on just those days when at least one of the flowers F_1 or F_2 is blue– namely, the flower $F_1 \cup F_2$. Also, for any three flowers F_1, F_2, and F_3, the flower $(F_1 \cup F_2) \cup F_3$ is easily seen to be blue on just those days when at least one of the three flowers F_1, F_2 or F_3 is blue. Similarly, for any four flowers F_1, F_2, F_3, F_4 the flower $((F_1 \cup F_2) \cup F_3) \cup F_4$ is blue on those and only those days on which at least one of these four flowers is blue. This generalizes to *any* finite number of flowers: Given any *finite* set S of flowers, there is a flower F that is blue on those and only those days when at least one of the flowers in S is blue. Furthermore, by the uniqueness condition, there cannot be more than one such flower. Thus, for any finite set of flowers, there is one and only one flower that is blue on those and only those days on which at least one flower in S is blue, and this flower we call the *join* of S (for reasons that will appear in a later chapter). Thus, for example, if S has 5 flowers F_1, F_2, F_3, F_4 and F_5, then the join of S is the flower $((((F_1 \cup F_2) \cup F_3) \cup F_4) \cup F_5)$–more briefly written $F_1 \cup F_2 \cup F_3 \cup F_4 \cup F_5$, or still more briefly $F_1 \cup \ldots \cup F_5$. [Incidentally, $F_1 \cup F_2$ is also called the *join* of F_1 with F_2, for reasons that will be given in a later chapter]. In general, if S has n flowers $F_1 \ldots F_n$, then the join of S is denoted $F_1 \cup \ldots \cup F_n$ (read F_1 join $\ldots$ join F_n).

Of course, the join of S dominates each flower in S, but it might dominate other flowers as well. However, we have a vital fact indicated in the next problem.

Problem 10 — Prove that if S is a *basic*, set, then the join of S dominates no basic flowers other than those in S, and hence the join of S represents S.

We now see that every basic set S *does* have a representative–namely the join of S. Also, a basic set cannot have more than one representative (Proposition B) and so we now have the following key fact.

Proposition C — A basic set S has one and only one representative–namely, the join of S.

Of course, each flower F represents one and only one basic set–namely the set F^* of all basic flowers dominated by F. Also, by Proposition C, a basic set has one and only one representative, and so we have:

Theorem 1 — In a reduced Boolean garden, each flower represents one and only one basic set, and each basic set is represented by one and only one flower.

1-1 Correspondences — A relation between a set S_1 and a set S_2 is called a *1-1 correspondence* between S_1, and S_2 if and only if the following two conditions hold:

(1) Each element of S_1 stands in the relation to one and only one element of S_2.

(2) For each element of S_2, there is one and only one element of S_1 that stands in the relation to it.

For example, in a monogamous country, the relation "x is the husband of y" is a 1-1 correspondence between the set M of married men of the country and the set W of married women in the country (for each x in M there is one and only one y in W such that x is the husband of y, and for each y in W, there is one and only one x in M such that x is the husband of y).

Another example: If in a theatre, all seats are taken and no one is standing, then the relation "person x is sitting on seat y" is a 1-1 correspondence between the set of people in the theatre and the set of seats in the theatre.

Obviously, if there is a 1-1 correspondence between one set and another, the two sets must have the same number of elements. For example, if you look into a theatre and see that all the seats are taken and no one is standing then, without having to count either the number of seats or the number of people, you know that the two numbers must be the same. Or again, in a monogamous country, the number of married men must be the same as the number of married women.

Now, Theorem 1 tells us that the relation "Flower F represents basic set S" is a 1-1 correspondence between the set of flowers and the set of basic sets, and so as an immediate corollary we have:

Theorem 2 — In a reduced Boolean garden, the number of flowers is the same as the number of basic sets.

Now we can solve the "grand" problem.

Problem 11 — Given that George's reduced Boolean garden has somewhere between 200 and 500 flowers in it, exactly how many flowers does it have? Also, how many of the flowers are basic?

SOLUTIONS

1 – These three conditions are all equivalent; *each* of them implies the other two. They really all say the same thing. We will show that (1) implies (2), and (2) implies (3), and (3) implies (1), which will close the circle.

Suppose (1)–F dominates G. Then on any day on which G is blue, F is also blue, hence $G \cap F$ is blue. Thus $G \cap F$ dominates G. Also, of course, G dominates $F \cap G$ (Fact (g)), and so G and $F \cap G$ are blue on the same days, and so $G \cap F = G$. Thus (1) implies (2).

Now suppose (2)–$G \cap F = G$. Then, on any day that G is blue, $G \cap F$ is blue, hence F is blue on such a day; hence $\overline{F}$ must then be red. And so there is no day on which G and $\overline{F}$ are both blue, hence $\overline{F} \cap G$ is never blue, and so $\overline{F} \cap G = r$. Thus (2) implies (3).

Now suppose (3)–$\overline{F} \cap G = r$. Consider any day on which G is blue. If F were red on that day, then $\overline{F}$ would be blue, hence $\overline{F} \cap G$ would be blue, which is not possible since $\overline{F} \cap G = r$. Thus F must be blue on any day on which G is blue, and so F dominates G. Thus (3) implies (1), which closes the circle. We thus have the following proposition [We recall that we write "iff" to abbreviate "if and only if"].

Proposition 1 — F dominates G iff $G \cap F = G$, which in turn is the case iff $\overline{F} \cap G = r$.

2 – Suppose that A is a flower dominated by both F and $\overline{F}$. Then whenever A is blue, F and $\overline{F}$ are both blue, which never happens. Thus A is always red, hence $A = r$. Hence we have.

Proposition 2 — The only flower that can be dominated by both F and $\overline{F}$ is r.

3 – Certainly not! Suppose X is basic and X dominates Y, where $Y \neq X$. Then Y must be r, hence Y is not basic. Therefore, X cannot dominate any basic flower Y which is different from X. We thus have.

Proposition 3 — No basic flower can dominate any other basic flower.

4 – By Proposition 1, F dominates X iff $X \cap F = X$, hence F fails to dominate X iff $X \cap F \neq X$. Now suppose that X is basic and F fails to dominate X. Then $X \cap F \neq X$. Yet X dominates $X \cap F$ (Fact (g)), and since $X \neq X \cap F$, then $X \cap F$ must be r. We thus have.

Proposition 4 — If F fails to dominate X, and X is basic, then $X \cap F$ is red on all days $(X \cap F = r)$.

5 – Suppose X is basic and F fails to dominate X. Then by Proposition 4, $X \cap F$ is always red. Hence on any day that X is blue, F must be red (for if F were then blue, $X \cap F$ would then be blue, which it never is), and hence $\overline{F}$ must be blue. Thus $\overline{F}$ is blue on all days that X is blue, and so $\overline{F}$ dominates X. And so, if F fails to dominate X, then $\overline{F}$ dominates X. We thus have.

Proposition 5 — For any basic flower X and any flower F, either F dominates X or $\overline{F}$ dominates X.

6 – Suppose X and Y are both basic and distinct. Then by Proposition 3, X cannot dominate Y, and so by Proposition 5, $\overline{X}$ dominates Y. Thus $\overline{X}$ is blue on all days on which Y is blue, hence X is red on all days on which Y is blue. Thus we have.

Proposition 6 — Two distinct basic flowers X and Y can never both be blue on the same day.

7 – Suppose X is basic. Then $X \neq r$, hence there is at least one day on which X is blue. No other basic flower can be blue on that day (by Proposition 6), and so all the other basic flowers must be red on that day. Thus it is (b) that is true, which we record as.

Proposition 7 — For any basic flower X, there is some day on which X is blue and all the other basic flowers are red.

8 – Consider any flower F_1, other than r. If F_1 is itself basic, then obviously F_1, dominates a basic flower–namely, F_1 itself. If F_1 is not basic, then F_1 dominates some flower F_2 other than F_1 and r. If F_2 is basic, then F_1 dominates the basic flower F_2. If F_2 is not basic, then it dominates some flower F_3 other than F_2 and r. Also $F_3 \neq F_1$, or we would have a domination cycle, contrary to Fact (h). If F_3 is not basic, then it dominates some new flower F_4 other than r (it must be new–that is, different from F_1, F_2 and F_3–to avoid a domination cycle), and so we keep going at this rate until we must finally reach some flower F_n such that the only new flower it can dominate is r itself (since there are only finitely many flowers in the garden, and at each step, a new flower is chosen). Then F_n is basic, and F_1 must dominate F_n (by repeated applications of Fact e).

We have thus proved.

Proposition 8 — Every flower other than r dominates some basic flower.

9 – Suppose F and G are distinct flowers. Then it cannot be that each of them dominates the other (for if they both did, then F would be the same flower as G by Fact f), hence one of them fails to dominate the other. Suppose that it is F that fails to dominate G. Then there is some day on which G is blue and F is red, hence on such a day, G and $\overline{F}$ are both blue, hence $\overline{F} \cap G$ is blue on that day, hence $\overline{F} \cap G \neq r$. Then by Proposition 8, $\overline{F} \cap G$ dominates some basic flower X. Then $\overline{F}$ and G both dominate X (since they both dominate $\overline{F} \cap G$, which in turn dominates X). Since $\overline{F}$ dominates X and $X \neq r$ (X is basic) then F doesn't dominate X (Proposition 2). Thus G dominates a basic flower that F does not dominate. [Of course, if it is G that fails to dominate F, then F dominates some basic flower that G doesn't dominate].

We thus have:

Proposition 9 — If F and G are distinct flowers, then one of them dominates some basic flower not dominated by the other.

10 – Suppose that S is a basic set. Let F be the join of S. Now let X be any basic flower not in S. We are to show that F fails to dominate X.

By Proposition 7, there is some day on which X is blue and all the other basic flowers are red, hence all the flowers in S are red, hence their join F must be red on such a day. Thus, there is a day on which X is blue and F is not, which means that F does *not* dominate X. Thus F dominates all and *only* those basic flowers that are in S, and so $S = F^*$. we thus have.

Proposition 10 — Every basic set S has a representative–namely the join of S.

11 – As is well known, the number of ways of selecting a set of objects from among n given objects is 2^n. For example, suppose you have a group of 5 children and you decide to take some of them (maybe all, maybe none) on a picnic. How many ways can this be done? Well, number the children in some order (maybe according to

age). For the first child you have 2 choices–to take him or not to take him. With each of these 2 choices, you have 2 choices for the second child; so for the first two children you have 4 choices altogether (either the first and not the second, or the second but not the first, or both, or neither). With each of these 4 choices for the first two children, you have 2 choices for the third–making 8 possible choices for the first three. With each of these 8, you have 2 choices for the fourth child, making 16 for the first four. Then with the fifth child, there are 2 possible choices, making 32 possible choices among the possible groups of 5 children. With 6 children there are 64 possible groups (2^6); with 7, there are 128 (2^7), and in general, with n objects (children or other things) the number of ways of forming a set of them (taking some, maybe all, maybe none) is 2^n.

Coming back to our flower garden, if n is the number of basic flowers, then 2^n is the number of basic sets, and by Theorem 2, this is also the total number of flowers! And so as a corollary of Theorem 2 we have:

Proposition 11 — In a finite reduced Boolean garden, the number of flowers must always be a power of 2–more specifically, 2^n, where n is the number of basic flowers.

In particular, for George's garden in which the number of flowers is somewhere between 200 and 500, the only power of 2 in this interval is 256, which is 2^8. And so George's garden has 8 basic flowers and 256 flowers all told.

Chapter 16

Boolean Gardens and Variable Liars

I – More on Boolean Gardens

We now need to know a little more about reduced Boolean gardens than what we found out in the last chapter.

We know that each flower F represents one and only one basic set–the set F^* of all basic flowers dominated by F. Also, F is the join of the set F^*. What basic set is represented by the flower b which is always blue? Obviously, b dominates *every* flower, and so b^* is thus the set of *all* the basic flowers of the garden. Thus also, b is the join of all the basic flowers of the garden.

What is the set r^* (r is the flower that is constantly red)? Well, r doesn't dominate any flower other than r, and no basic flower is r, and so r doesn't dominate any basic flower at all. Hence the set r^* represented by r consists of no flowers at all; it is what is known as the *empty set*.

A set is called *empty* if it contains no members at all, such, for example, as the set of all people in a theatre after everyone has left. Two sets are called *identical* if and only if they have exactly the same members–in other words, neither set contains a member that is not also in the other one. Well, if E_1 and E_2 are both empty sets, then surely neither one contains a member that is not in the other, since neither one contains any member at all. Thus E_1 and E_2 contain exactly the same members–namely, no members at all–and so $E_1 = E_2$. Thus, there is only one empty set, and it is denoted by the symbol "$\emptyset$". And so, we see that b represents the set of all basic flowers, and at the other extreme, r represents the set $\emptyset$ consisting of no flowers at all.

Now for some questions:

Problem 1 – What is the minimum number of flowers that a reduced Boolean garden can have? [We assume, of course, that there exists at least one day].

Problem 2 – Is it possible in a reduced Boolean garden that b can be a basic flower?

Problem 3 – In a reduced Boolean garden, can there be a day on which all the basic flowers are red?

Problem 4 – Suppose that X is a basic flower of a reduced Boolean garden and that X is blue on one day–call it Day 1–and is blue on another day–call it Day 2. Does it necessarily follow that every one of the flowers must be of the same color on Day 1 as on Day 2?

Problem 5 – Let us say that two days D_1 and D_2 are *similar*, or *of the same type* if each of the flowers is the same color on both days. The question is how many different types of days are there for George's garden? Put otherwise, how many possible color configurations are there in George's garden?

At this point, the reader should be familiar with the solutions to the above problems. It will be convenient to now review some of the things we have learned in the last chapter and so far in this chapter. We recall that a flower F is said to *dominate* flower G if F is blue on all days on which G is blue, and a flower is called *basic* if it is not r and dominates no flower other than itself and r. F^* is the set of all *basic* flowers dominated by F, and also F is the *join* of F^*–the flower that is blue on all and only those days on which at least one flower of the set F^* is blue. We also recall the following facts:

F_1: F dominates G if and only if $F \cap G = F$. Also F dominates G if and only if $\overline{F} \cap G = r$.

F_2: For any basic flower X, either F or $\overline{F}$ dominate X, but not both.

F_3: On each day exactly one basic flower is blue and all the other basic flowers are red. Also, for each basic flower X, there is a day on which X is blue and all the other basic flowers are red.

F_4: Which basic flower is blue on a given day determines the entire color configuration of the garden on that day. Thus if Γ is blue on one day on which basic flower X is blue, then F is blue on every day on which X is blue, and so F then dominates X. [Also, if F is red on one day on which X is blue, then F is red on all days that X is blue].

F_5: The set of flowers is in a 1-1 correspondence with the set of sets of basic flowers–each flower F corresponding to the set F^* of flowers dominated by F. Also, F is the *join* of F^*–that is, F is blue when and only when at least one flower in F^* is blue. Thus, the number of flowers is the same as the number of basic sets, this number being 2^n, where n is the number of basic flowers.

We now need to establish another fact: For any flowers A, B and C, it is obvious that $A \cap B$ dominates C if and only if A and B each dominate C. It may be tempting to believe that $A \cup B$ dominates C if and only if at least one of A or B dominates C, but in general this is not true! Of course, if either A or B dominates C, then so does $A \cup B$, since $A \cup B$ dominates both A and B. But if $A \cup B$ dominates C, it does not follow that either A dominates C or B dominates C. For example, consider a Boolean garden with more than two flowers. Let A be any flower other than b or r. Then A is also different from both b and r. Neither A nor $\overline{A}$ dominates b (b is the only flower that dominates b), yet $A \cup \overline{A}$ does dominate b, since $A \cup \overline{A}$ *is* the flower b. And so in general, it is not true that if $A \cup B$ dominates C, then either A or B dominate C, but we do have the following important fact:

F_6: For any *basic* flower X, if $A \cup B$ dominates X, then at least one of the flowers A or B must dominate X.

Problem 6 – Prove F_6.

Sets of Basic Flowers – For any set S of basic flowers, by S' we shall mean the set of all basic flowers that are *not* in S. [We call S' the *complement* of S (relative to the set of basic flowers)]. For any sets S_1 and S_2 of basic flowers, by $S_1 \cup S_2$ is meant the set consisting of all the flowers of S_1 together with all the flowers in S_2. Thus, $S_1 \cup S_2$ is the set of all basic flowers that belong either to S_1 or to S_2, or both. [We call $S_1 \cup S_2$ the *union* of S_1 with S_2]. By $S_1 \cap S_2$ is meant the set of all basic flowers that belong to both S_1 and S_2. [We call $S_1 \cap S_2$ the *intersection* of S_1 with S_2].

As an example, suppose there are eight basic flowers A, B, C, D, E, F, G, H in the garden. Suppose S_1 is the set {A, B, C, D, E} and S_2 is the set {D, E, F, G}. Then $\overline{S_1} = \{F, G, H\}$; $\overline{S_2} = \{A, B, C, H\}$; $S_1 \cup S_2 = \{A, B, C, D, E, F, G\}$ and $S_1 \cap S_2 = \{D, E\}$. Also $\overline{S_1 \cup S_2} = \{H\}$ (the set consisting of just the one flower H).

Now comes an important problem.

Problem 7 – Prove fact F_7 below.

F_7 – For any flowers F and G:

(1) $\overline{F}^* = F^{*'}$. [In other words the set of basic flowers dominated by $\overline{F}$ is the complement of the set of basic flowers dominated by F].

(2) $(F \cup G)^* = F^* \cup G^*$. [In other words, the set of basic flowers dominated by $F \cup G$ is the *union* of the set of flowers dominated by F with the set of flowers dominated by G].

(3) $(F \cap G)^* = F^* \cap G^*$. [In other words, the set of flowers dominated by $F \cap G$ is the *intersection* of the set of flowers dominated by F with the set of flowers dominated by G].

II – Boolean Gardens and Boolean Islands

We recall that in Chapter 12 we dealt with *variable liars*–people that may lie on some days and be truthful on others. We now consider an island in which all the inhabitants are variable liars. The behavior of each inhabitant is constant throughout any given day (each one either lies all day, or is truthful all day), but his or her behavior can change from one day to another. We are given that for any inhabitants A and B, there is an inhabitant C who lies on those and only those days on which A and B both tell the truth (Just as in a Boolean garden, for any flowers A and B, there is a flower C which is red on those and only those days on which A and B are both blue). Such an island we will call a *Boolean island*. This kind of island is exactly like a Boolean garden, except that flowers are replaced by people, and "blue" by "truthful" and "red" by "lying". We call two inhabitants *similar* if they are truthful on just the same days, and we call the island a *reduced* Boolean island if no two distinct inhabitants are similar. We will now assume that our island is a *reduced* Boolean island. Then, as with Boolean gardens, we let $\overline{A}$ be the individual who is truthful on those and only those days when A lies, and for any inhabitants A and B, we let $A \cap B$ be that inhabitant who is truthful on those and only those days on which A and B are both truthful, and $A \cup B$ be that inhabitant who is truthful on those and only those days on which at least one of A and B is truthful. We let L be the constant liar–the one who lies on all days ($L = A \cap \overline{A}$ for any inhabitant A). In analogy with flower gardens, we say that A *dominates* B if A is truthful on all days on which B is truthful, and we call an inhabitant X *basic* if X dominates no inhabitant other than X and L, and $X \neq L$.

Everything we have proved so far about reduced Boolean gardens goes over intact to reduced Boolean islands, replacing "blue" by "truthful" and "red" by "lying". In particular, facts $F_1 - F_7$ for flower gardens have their exact analogies $I_1 - I_7$ below for Boolean islands.

I_1: Inhabitant A dominates inhabitant B if and only if $A \cap B = A$. Also A dominates B if and only if $\overline{A} \cap B = L$.

I_2: For any basic inhabitant X, either A or $\overline{A}$ dominates X, but not both.

I_3: On each day, exactly one basic inhabitant is truthful and all the other basic inhabitants lie. Also, for each basic inhabitant X, there is a day on which X is truthful and all the other basic inhabitants lie.

I_4: For any basic inhabitant X and any inhabitant A, if A is truthful (lying) on one day on which X is truthful, then A is respectively truthful (lying) on every day on which X is truthful.

I_5: The set of inhabitants is in 1-1 correspondence with the set of sets of basic inhabitants–each inhabitant A corresponding to the set A^* of basic inhabitants dominated by A. Also A is the *join* of A^* in the sense that A is truthful on those and only those days on which at least one member of the set A^* is truthful. The number of inhabitants is the same as the number of sets of basic inhabitants, which is 2^n, where n is the number of basic inhabitants.

I_6: For any *basic* inhabitant X, if $A \cup B$ dominates X, then at least one of the inhabitants A or B must dominate X.

I_7: For any inhabitants A and B:

(1) $\overline{A}^* = A^{*'}$. [$A^{*'}$ means the set of all basic inhabitants not in A^*].

(2) $(A \cup B)^* = A^* \cup B^*$.

(3) $(A \cap B)^* = A^* \cap B^*$.

Now, it so happens that in some remote region of some far-off ocean, there are two neighboring reduced Boolean islands, one of which is inhabited all by unmarried males, and the other, all by unmarried females, and the number of males happens to be the same as the number of females. The males would like to marry the females in such a way that for any males M_1 and M_2 and their corresponding wives W_1 and W_2, the following three conditiona hold:

(1) $\overline{M}_1$ is married to $\overline{W}_1$.

(2) $M_1 \cap M_2$ is married to $W_1 \cap W_2$.

(3) $M_1 \cup M_2$ is married to $W_1 \cup W_2$.

Such a marriage arrangement, if possible, we will call a *Boolean match*. The first question now is whether a Boolean match is always possible, regardless of the number of inhabitants. The second question is whether the number of possible Boolean matches can be determined just from knowing the number of inhabitants. I'll tell you right now that the answer to both questions is *yes*. I'll even tell you that if the number of inhabitants on each island is 256 (which is the number of flowers in George B.'s reduced garden), then the number of possible Boolean matches is 40,320.

If the reader is completely baffled by how the number 40,320 was ever obtained, I'll be delighted! I'll also reassure such a reader that he or she will become pleasantly unbaffled after reading the solutions.

Now comes a second problem: In the middle of the men's island, there happens to be a reduced Boolean flower garden, which coincidentally happens to have the same number of flowers as there are men. By an even stranger coincidence, the set of men is *synchronized* with the set of flowers, in the sense that for any two days on which each man behaves alike (each one either lies on both days or tells the truth on both days) each flower is the same color on both days. One day, the men decided that they would like to arrange matters so that each one picks and wears a flower in his lapel in such a way that each man's flower was blue on those days on which he told the truth, and red on those days on which he lied. [If this could be done, it would be of great advantage to visitors to the island, for a visitor could then tell of any given man whether he was truthful or lying on that day, by simply observing the color of the flower he was wearing]. The first question is whether this scheme can be carried out. The second question is how many ways can the scheme be carried out, assuming that the number of men is 256.

Another problem: On the women's Boolean island, there is also a Boolean flower garden with the same number of flowers as there are women and the women are synchronized with the flowers. One day the women decided on a more clever scheme than that of the men–a scheme that would fool all future visitors to the island! The scheme was that each woman should pick and wear in her hair a flower which would be *red* on those days on which she was truthful, and *blue* on those days on which she lied. Can this scheme be carried out?

III – Solutions to the Three Problems

First, we will solve the first problem through a little series of problems.

Problem 8 – Suppose that under a given Boolean match, males M_1, M_2 and M_3 are married to W_1, W_2 and W_3 respectively. Prove the following facts:

(1) $\overline{M}_1 = \overline{M}_2$ if and only if $\overline{W}_1 = \overline{W}_2$.

(2) $M_1 \cup M_2 = M_3$ if and only if $W_1 \cup W_2 = W_3$.

(3) $M_1 \cap M_2 = M_3$ if and only if $W_1 \cap W_2 = W_3$.

Problem 9 – Suppose that in a Boolean match, males M_1 and M_2 have respective wives W_1 and W_2. Prove that M_1 dominates M_2 if and only if W_1 dominates W_2.

Problem 10 – Prove that in a Boolean match the constant male truth teller must be married to the constant female truth teller, and the constant male liar must be married to the constant female liar.

Problem 11 – Prove that in a Boolean match, each basic male must be married to a basic female.

Problem 12 – Prove that however way we marry the basic males to the basic females, there is then one and only one way to marry all the others to achieve a Boolean match.

Problem 13 – Now we can determine the number of possible Boolean matches, given the number of couples. Suppose there are 128 males. How many possible Boolean matches are there? What if the number is 1024? What if the number is 256?

Problem 14 – Now let's consider the problem of the men picking the flowers. Show that the number of ways does not depend on the number of inhabitants, and determine what this number is.

Problem 15 – Is the women's scheme possible?

SOLUTIONS

1 – Since b is blue on all days and r is red on all days and there is at least one day, then there is at least one day on which b is blue and r is red, hence $b \neq r$. Thus, every Boolean garden must contain at least two flowers (b and r). Rut also, the set {b, r} consisting of just the flowers b and r forms a tiny Boolean garden in its own right (as the reader can easily verify), and this little garden is obviously reduced. And so, 2 is the minimum possible number of flowers in a reduced Boolean garden.

2 – Yes, but only if b and r are the only flowers in the garden, for then b dominates only itself and r, hence b is basic. Rut if the garden contains more than two flowers, then b must dominate some flower F other than b or r (since all the flowers are dominated by b), hence b is then not basic.

So the complete answer is that in a 2-element Boolean garden, b is basic, but if the garden has more than two flowers, then b is not basic.

3 – The flower b is the join of the set of all the basic flowers (since b represents this set), and so if all the basic flowers were red an some day, then their join would be red on that day, which would mean that b is red on that day, which is not possible. Therefore, there is no day on which all the basic flowers are red.

This fact, together with the previously proven fact that no two distinct basic flowers can ever be blue on the same day leads us to the realization that on each day, one basic flower is blue and all the other basic flowers are red.

4 – Yes, it does! X is blue on Day 1 and Day 2, and on both days, all the other basic flowers are red. Now each flower F is the join of one and only one basic set–the set F^*. Well, F^* either contains X or it doesn't. If it does, then F is blue on both Day 1 and Day 2 and if it doesn't, then F is red on both days. Thus, F is the same color on both days.

And so, we see that on each day, the entire color distribution of all the flowers is completely determined by which one of the basic flowers happens to be blue on that day!

5 – We have just seen that the entire color configuration of the garden on a given day is completely determined by which one of the basic flowers happens to be blue on that day. Thus the number of possible color configurations is the same as the number of basic flowers. For George's garden, this number is eight.

6 – Suppose X is basic and that $A \cup B$ dominates X. On some day, X is blue, and $A \cup B$ is therefore also blue on that day. Hence on that day, either A or B (maybe both) must be blue. Whichever one is blue on that day must be blue on *all* days that X is blue (by F_4, since X is basic) and hence must dominate X.

7 – To show that two sets S_1, S_2 of flowers are the same, it suffices to show that every flower in S_1 is in S_2, and that every flower in S_2 is in S_1. If this can be

shown, then S_1 and S_2 contain exactly the same flowers, and hence must be the same set. In what follows, X is any *basic* flower.

(1) Suppose flower X is in $\overline{F}^*$. Then $\overline{F}$ dominates X. Then F doesn't dominate X (by F_2). Hence X is not in the set F^*, so X is in its complement $F^{*'}$. Thus $F^{*'}$ contains all the flowers in $\overline{F}^*$.

Conversely, suppose X is in $F^{*'}$. Then X is not in F^* hence F doesn't dominate X, and therefore $\overline{F}$ does dominate X (by fact F_2), and so X is in $\overline{F}^*$. Thus, $\overline{F}^*$ contains all the flowers in $F^{*'}$, which concludes the proof that $\overline{F}^* = F^{*'}$.

(2) Suppose X is in $F^* \cup G^*$. Then X is in either F^* or G^* (maybe both), hence either F or G dominates X, and in either case, $F \cup G$ dominates X (because $F \cup G$ dominates both F and G), which means that X is in $(F \cup G)^*$. Thus all flowers in $F^* \cup G^*$ are in $(F \cup G)^*$.

Conversely, suppose X is in $(F \cup G)^*$. Then X is dominated by $F \cup G$, and X is basic. Therefore, by fact F_6, either F or G must dominate X, and so X is in either F^* or G^*, and in either case it is in $F^* \cup G^*$. Thus, all flowers in $(F \cup G)^*$ are in $F^* \cup G^*$, which concludes the proof that $(F \cup G)^* = F^* \cup G^*$.

(3) The statement $(F \cap G)^* = F^* \cap G^*$ says nothing more nor less than that a basic flower X is dominated by $F \cap G$ if and only if it is dominated by both F and G, which is obvious. [In fact *any* flower H, basic or not, is dominated by $F \cap G$ if and only if it is dominated both by F and by G].

8 – This is really so obvious that it is questionable whether I should have given it as a problem. I will give just the solution of (2), since the proofs for (3) and (1) are similar.

$M_1 \cup M_2$ is married to $W_1 \cup W_2$ and M_3 is married to W_3. Now, if $M_1 \cup M_2$ is the same person as M_3, then their wives are the same person, and thus $W_1 \cup W_2 = W_3$. Conversely, if $W_1 \cup W_2 = W_3$, then the husband of $W_1 \cup W_2$ is the same person as the husband of W_3, which means that $M_1 \cup M_2 = M_3$.

9 – Suppose M_1 dominates M_2. Then $M_1 \cap M_2 = M_1$ (by fact I_1). Hence $W_1 \cap W_2 = W_1$ (Problem 8), and hence W_1 dominates W_2 (fact I_1). Conversely, if W_1 dominates W_2, then $W_1 \cap W_2 = W_1$, hence $M_1 \cap M_2 = M_1$, and so M_1 dominates M_2.

10 – Let M be any male and W his wife (under a Boolean match). Then $\overline{M}$ must be married to $\overline{W}$, and hence $M \cup \overline{M}$ must be married to $W \cup \overline{W}$, but $M \cup \overline{M}$ is the constant male truth-teller and $W \cup \overline{W}$ is the constant female truth-teller, and so those two are married. Also, $M \cap \overline{M}$ must be married to $W \cap \overline{W}$, and thus the constant male liar is married to the constant female liar.

11 – Let L_1 be the constant male liar and L_2 be the constant female liar. Then L_1 is married to L_2, as we saw above. Now, let M be any male and W his wife. If M is basic, then M dominates only M and L_1, hence W dominates only W and L_2 (Problem 9), which means that W is basic. Conversely, if W is basic, then W

dominates only W and L_2, hence M dominates only M and L_1, and thus M is basic. This proves that M is basic if and only if W is basic.

12 – Once the basic males are married to the basic females, each male M marries according to the following scheme: M^* is the set of basic men dominated by M. Let S be the set of wives of the men in M^*. Then there is one and only one woman W such that $W^* = S$ (namely, W is the *join* of S). Then M marries W. And so, for any male M and female W, if W^* is the set of wives of the men in M^*, then M is married to W. Conversely, if M is married to W, then W^* must be the set of wives of the men in M^* for let S be this set of wives. Then M is married to the join of S, and since M is married to W, then W is the join of S, and hence S is the set W^*. Thus W^* is the set of wives of the men in M^*.

We thus have the following basic fact: For any man M and woman W, *M is married to W if and only if W^* is the set of wives of the men in M^**.

Now we show that this marriage is a Boolean match.

(1) Suppose M is married to W. We are to show that $\overline{M}$ is married to $\overline{W}$. Well, since M is married to W, then W^* is the set of wives of the men in M^*. Then each basic woman *not* in W^* is the wife of some basic man *not* in M^*, and each basic man *not* in M^* is married to a basic woman *not* in W^*, and thus the set of basic women not in W^* is the set of wives of the basic men who are not in M^*. Thus $\overline{W^*}$ is the set of wives of the men in $\overline{M^*}$, but by fact I_7, $\overline{W^*} = \overline{W}^*$ and $\overline{M^*} = \overline{M}^*$. Therefore, $\overline{W}^*$ is the set of wives of the men in $\overline{M}^*$, and so $\overline{M}$ is married to $\overline{W}$, which was to be proved.

(2), (3) Suppose M_1 is married to W_1 and M_2 to W_2. We are to show that $M_1 \cap M_2$ is married to $W_1 \cap W_2$, and that $M_1 \cup M_2$ is married to $W_1 \cup W_2$. Now, W_1^* is the set of wives of the men in M_1^*, and W_2^* is the set of wives of the men in M_2^*, and therefore $W_1^* \cap W_2^*$ is the set of wives of the men in $M_1^* \cap M_2^*$ (because every woman in $W_1^* \cap W_2^*$ is in both W_1^* and W_2^*, hence is married to a man who is in both M_1^* and M_2^*, and hence is in $M_1^* \cap M_2^*$, and similarly, every man in $M_1^* \cap M_2^*$ must be married to a woman in $W_1^* \cap W_2^*$). However, $M_1^* \cap M_2^* = (M_1 \cap M_2)^*$ and $W_1^* \cap W_2^* = (W_1 \cap W_2)^*$ (by fact I_7) and so $(W_1 \cap W_2)^*$ is the set of wives of the men in $(M_1 \cap M_2)^*$. Therefore, $M_1 \cap M_2$ must be married to $W_1 \cap W_2$.

The proof that $M_1 \cup M_2$ is married to $W_1 \cup W_2$ is along similar lines. [Incidentally, condition (3) of the definition of a Boolean match was not necessary to explicitly state, since it is derivable from the other two conditions, using the fact that $A \cup B = (\overline{A \cap B})$, which does hold for any reduced Boolean setup, whether flower gardens or islands].

13 – We have seen that under any Boolean match, basic males must marry basic females. Also, we have just seen that *any* marriage between basic males and basic females can be extended in one and only one way to a Boolean match. And so, the number of possible Boolean matches is the same as the number of ways that the basic males can marry the basic females. How many such ways are there, given the number of inhabitants of each island?

Well, suppose for example that there are 64 males and 64 females. Now, $64 =$

2^6, and so there are then 6 basic males and 6 basic females. Arrange the 6 basic males in any order. For the first male, there are 6 possibilities for his wife. Once chosen, there remain 5 possibilities for the second male, hence there are 6x5 = 30 possibilities for the first two males. With each of these 30 possibilities, there are 4 possibilities for the third male, and so there are 30x4 = 120 possibilities for the first three males. With each of these possibilities there are 3 possibilities for the fourth male, making 120x3 = 360 possibilities for the first four males. With each of these, there are 2 possibilities for the fifth male, making 2x360 = 720 possibilities for the first 5 males. Then, there is only one possibility left for the sixth male, and so there are 720 possibilities all told. Thus, if there are 6 basic males and 6 basic females, the number of possible Boolean matches is 6x5x4x3x2x1 = 720. If there were 7 basic males and 7 basic females (which is the case if there are 256 males and 256 females), then the answer would be 7x6x5x4x3x2x1, which is 7x720 = 5040. If there were 8 basic males and 8 basic females (which is the case if there are 256 males and 256 females), then the answer would be 8x7x6x5x4x3x2x1 = 8x5040 = 40, 320. [Now you see how the number 40,320 was obtained!]

In general, for any positive whole number n by *factorial n* (sometimes called *n factorial*), written n!, is meant the product of all the positive whole numbers from 1 to n. And so in Boolean islands of 2^n males and 2^n females, there are n basic males and n basic females, and the number of possible Boolean matches is n!. If the number of inhabitants of each island is 1024, then $1024 = 2^{10}$, and so there are 10 basic males and 10 basic females, so that number of possible Boolean matches is then 10!–i.e. 10x9x8x7x6x5x4x3x2x1, which is 10x9x(8x7x6x5x4x3x2x1) = 10x9x40, 320 = 90x40, 320 = 3,628,800.

14 – This can be done in one and only one way, and the way is this: Each basic man is truthful on some day and an this day, one and only one basic flower is blue. He then picks and wears this basic flower. [The particular day he is truthful and picks his basic flower doesn't matter, since by the given synchronization condition, if a basic flower is blue on one day on which a basic man is truthful, it is blue on *all* days on which he is truthful]. Thus, there is one and only one way in which the basic men can pick their basic flowers. Once this is done, then, in similarity to the first problem all the men pick their flowers according to the rule: Man M picks flower F if and only if F^* is the set of flowers picked by the men in M^*. Then on any day, M is truthful if and only if one of the men in M^* is truthful, which is the case if and only if one of the flowers in F^* is blue, which in turn is so if and only if F is blue.

We thus see that the number of inhabitants and the number of flowers has nothing to do with it (assuming, of course, that the numbers are the same). Regardless of this number, there is one and only one way the males can pick their flowers to match their truthfulness (blue matching truth and red matching falsehood).

15 – Yes, it is possible. The simplest way to see this is the following: Each woman picks and temporarily holds a flower in the same way as in the case of the men, so that the flower she holds is blue on all days on which she is truthful, and red on

all days on which she lies. Then each woman W simply exchanges her flower with that of $\overline{W}$.

Chapter 17

Propositional Logic and Boolean Gardens

All the things we have proved in the last four chapters are but special cases of various results in the field known as *Boolean Algebra*, a subject that is of basic importance today, not only in the foundations of mathematics, but also in the practical fields of electrical engineering and computer science. We will study Boolean algebra in its complete generality in Chapters 19, 20 and 21. In this chapter we turn to the related field of *propositional logic* and its relation to Boolean gardens, and in the next chapter we study the Boolean algebra of *sets* (a special case of Boolean algebra in general).

Remarks – Someone once asked me: "What is this strange subject of Boolean algebra with such weird equations as $1 = 1 = 0$? I replied that in Boolean algebra, the symbols 1, +, and 0 have entirely different meanings than in ordinary arithmetic. What these meanings are will be explained in a later chapter. But now, let us turn to the basic field of propositional logic.

I – Propositional Logic

The Logical Connectives – Just as in ordinary algebra, we use letters x, y, z with or without subscripts, as standing for arbitrary numbers, so in propositional logic we use letters p, q, r with or without subscripts as standing for arbitrary *propositions*.

Propositions can be combined by using the so-called *logical connectives*. The principal ones are:

(1) ′ not

(2) ∧ (and)

(3) $\vee$ (or)

(4) $\supset$ (if-then)

(5) $\equiv$ (if and only if)

Here is what they mean:

(1) **Negation** – For any proposition p, by p' (sometimes written $\sim$ p, or sometimes $\neg$ p) is meant the *opposite* or *contrary* of p. For example, if p is the proposition that Jack is guilty, then p' is the proposition that Jack is *not* guilty. The proposition p' is read "it is not the case that p", or more briefly, "not p". The proposition p' is called the *negation* of p, and is true if p is false, and false if p is true. These two facts are summarized in the following table, which is called the *truth table for negation*. In this table, as in all the tables that follow, we shall use the letter "T" to stand for *truth* and "F" to stand for *falsehood*.

p	p'
T	F
F	T

The first line of this truth table says that if p has the value T (in other words, if p is true), then p' has the value, F (p' is false). The second line says that if p has the value F, then p' has the value T. We can also express this in the following equations:

$$T' = F$$
$$F' = T$$

(2) **Conjunction** – For any propositions p and q, the proposition that p *and* q are both true is written "$p \wedge q$" (sometimes "p & q"). We call $p \wedge q$ the *conjunction* of p and q, and it is read "p and q are both true", or more briefly, "p and q". For example, if p is the proposition that Jack is guilty and q is the proposition that Jill is guilty, then $p \wedge q$ is the proposition that Jack and Jill are both guilty.

The proposition $p \wedge q$ is true if p and q are both true and is false if at least one of them is false. We thus have the following four laws of conjunction:

$$T \wedge T = T$$
$$T \wedge F = F$$
$$F \wedge T = F$$
$$F \wedge F = F$$

This is also expressed by the following table–the truth table for *conjunction*.

p	q	$p \wedge q$
T	T	T
T	F	F
F	T	F
F	F	F

(3) **Disjunction** – We write "$p \vee q$" to mean that at least one of the propositions p, q is true (and maybe both). We read "$p \vee q$" as "either p or q", or more briefly "p or q". It is true if at least one of the propositions p, q is true, and false only if p and q are both false.

For example, if p is the proposition that Jack is guilty and q is the proposition that Jill is guilty, then $p \vee q$ is the proposition that at least one of the two persons, Jack or Jill is guilty (and maybe both).

It should be pointed out that in ordinary English, the phrase "either-or" is used in two senses–the *strict* or *exclusive* sense, meaning *exactly one*, and the *loose* or *inclusive* sense, meaning *at least one*. For example, if I say that tomorrow I will marry either Betty or Jane, I of course mean that I will marry one *and only one* of the two ladies, and so I am then using "either-or" in the exclusive sense. On the other hand, if an advertisement for a secretary requires that the applicant know either French or German, an applicant is certainly not going to be rejected because she happens to know both! So in this case, "either-or" is used in the *inclusive* sense. Now, in formal logic, mathematics and computer science, we always use "either-or" in the *inclusive* sense, and so $p \vee q$ means that *at least one* of p, q is true.

I might also point out that in Latin, there are two different words for the two different senses: "aut" is used for the exclusive sense, and "vel" for the inclusive sense. In fact, the logical symbol "$\vee$" for "or" actually comes from the Latin word "vel".

The proposition $p \vee q$ is called the *disjunction* of p and q, and has the following truth table.

p	q	$p \vee q$
T	T	T
T	F	T
F	T	T
F	F	F

We noted that we have an F for $p \vee q$ only in the last row (in which p and q are both F). This table can also be expressed by the equations:

$$T \vee T = T$$
$$T \vee F = T$$
$$F \vee T = T$$
$$F \vee F = F$$

(4) **If-Then** – The if-then symbol "⊃" is particularly troublesome to those first contacting symbolic logic, since it is questionable whether the meaning of "if-then" as used technically by logicians is quite the same as that of common use.

For any proposition p and q, we write "$p \supset q$" to mean "if p, then q"; also read "p *implies* q", or "it is not the case that p is true and q is false", or "either p is false, or p and q are both true". Thus, $p \supset q$ means that you can't have p without also having q, or in other words that either p is false, or p and q are both true.

How are we to evaluate the truth or falsity of $p \supset q$, given the truth or falsity of each of p and q? Well, there are 4 cases to consider. Either p and q are both true, or p is true and q is false, or p is false and q is true, or p and q are both false. In the first case, since q is true, then certainly it is the case that *if* p, then q (because for that matter, if *not* p then q would also hold. If q is true absolutely, then "if p then q" is quite independent of p). And so we clearly have

(1) T ⊃ T = T

Next , if p is true and q is false, then $p \supset q$ must be false (because a true proposition can never imply a false proposition), and so we have

(2) T ⊃ F = F

In the third case, since q is true, then it is also the case that $p \supset q$ is true, regardless of whether p is true or false, and so we have

(3) F ⊃ T = T

Now, the fourth case is the puzzling one: Suppose p and q are both false; what should we make of "*if* p, *then* q? Some might guess that it should be false, others that it is true, and others that it is inapplicable. At that point, a decision must be made once and for all, and the decision that *has* been made by logicians and computer scientists is that in this case, $p \supset q$ should be declared *true*. Let me give you what I believe is a good argument why this decision is a wise one.

Suppose I put a card face down on the table, without first letting you see the face. I then say: "If this card is the queen of spades, then it is black". Surely, you will agree! Thus, letting p be the proposition that the card is the queen of spades and q the proposition that the card is black, I am asserting $p \supset q$, and you agree. Now, suppose I turn this card over, and it turns out to be the five of diamonds, am I then to retract my statement? You originally agreed that $p \supset q$ is true, but you have subsequently seen that p and q are both false (the card is neither the queen of spades nor black), but isn't it still true that *if* the card had been the queen of spades, *then* it would have been black? And so, here is a perfect example of a case where $p \supset q$ is true even though p and q are themselves both false. And so, we have

(4) F ⊃ F = T

Thus, the truth table for ⊃ is the following:

p	q	$p \supset q$
T	T	T
T	F	F
F	T	T
F	F	T

It must be emphasized that $p \supset q$ is false *only* in the one case that p is true and q is false–in the other three cases, it is true.

(5) **If-and-only-if** – We write $p \equiv q$ to mean that p and q are either both true or both false; or what is the same thing, if either one is true, so is the other. We read $p \equiv q$ as "p if and only if q" or "p and q are equivalent" (as far as their truth or falsity are concerned).

Since $p \equiv q$ is true when and only when p and q are either both true or both false, then $p \equiv q$ is false when and only when one of p, q is true and the other false (either p true and q false, or p false and q true), so here is the truth table for $\equiv$

p	q	$p \equiv q$
T	T	T
T	F	F
F	T	F
F	F	T

Or, as equations:

$$T \equiv T = T$$
$$T \equiv F = F$$
$$F \equiv T = F$$
$$F \equiv F = T$$

We note that $p \equiv q$ holds if and only if $p \supset q$ and $q \supset p$ both hold. Thus, $p \equiv q$ could be regarded as a shorthand for $(p \supset q) \wedge (q \supset p)$. We remark that the operation $\supset$ is sometimes called the *conditional*, and $\equiv$, the *bi-conditional*.

The operations $', \wedge, \vee, \supset, \equiv$ are examples of what are called *logical connectives*.

Parentheses – One can combine simple propositions into compound ones in many ways, using the logical connectives. We usually need *parentheses* to avoid ambiguity. For example, if we write $p \wedge q \vee r$ without parentheses, one cannot tell which of the following is meant:

(1) Either $p \wedge q$ is true, or r is true.

(2) p is true and $q \vee r$ is true.

If we mean (1), we should write $(p \wedge q) \vee r$, whereas if we mean (2), we should write $p \wedge (q \vee r)$. [The situation is analogous to algebra: $(x + y) \times z$ has a different meaning from $x + (y \times z)$–for example, $(2 + 3) \times 4 = 20$, whereas $2 + (3 \times 4) = 14$.] We thus need parentheses in propositional logic for punctuation.

Compound Truth Tables – By the *truth value* of a proposition p is meant its truth or falsity–that is T, if p is true, and F, if p is false. Thus the proposition that $2 + 3 = 5$ and Paris is the capital of France, though different propositions, have the same truth value, namely T.

Consider now two propositions p and q. If we know the truth value of p and the truth value of q, then by the simple truth tables already constructed, we can determine the truth values of $p', p \wedge q, p \vee q, p \supset q$ and $p \equiv q$. It therefore follows that given any combination of p and q–that is, any proposition expressible in terms of p and q, using the logical connectives, we can determine the truth value of this combination, given the truth values of p and q. For example, suppose X is the combination $(p \equiv (q \wedge p)) \supset (p' \supset q)$. Given the truth values of p and of q, we can successively find the truth values of $q \wedge p, p \equiv (q \wedge p), p', p' \supset q$. and finally $(p \equiv (q \wedge p)) \supset (p' \supset q)$. There are four possible distributions of truth values for p and q. (p true, q true; p true q false; p false, q true; and p false, q false), and in each of the four cases, we can determine the truth value of X. We can do this systematically by constructing the following table (an example of a *compound truth table*):

p	q	$q \wedge p$	$p \equiv (q \wedge p)$	p'	$p' \supset q$	$(p \equiv (q \wedge p)) \supset (p' \supset q)$
T	T	T	T	F	T	T
T	F	F	F	F	T	T
F	T	F	T	T	T	T
F	F	F	T	T	F	F

We see that X is true in the first three cases and false in the fourth.

We can also construct a truth table for a combination of three propositional unknowns–p, q and r–but now there are eight cases to consider (because there are four distributions of T's and F's to p and q, and with each of these four distributions there are two possibilities for r). For example, suppose X is the combination $(p \wedge q) \equiv (p' \supset r)$. Here is a truth table for X:

p	q	r	$p \wedge q$	p'	$p' \supset r$	$(p \wedge q) \equiv (p' \supset r)$
T	T	T	T	F	T	T
T	T	F	T	F	T	T
T	F	T	F	F	T	F
T	F	F	F	F	T	F
F	T	T	F	T	T	F
F	T	F	F	T	F	T
F	F	T	F	T	T	F
F	F	F	F	T	F	T

We see that X is true in Cases 1, 2, 6 and 8.

Tautologies – Consider now the following expression

$$(p \supset q) \equiv (q' \supset p')$$

Its truth table is the following:

p	q	p'	q'	$p \supset q$	$q' \supset p'$	$(p \supset q) \equiv (q' \supset p')$
T	T	F	F	T	T	T
T	F	F	T	F	F	T
F	T	T	F	T	T	T
F	F	T	T	T	T	T

We notice that the last column contains all T's. Thus $(p \supset q) \equiv (q' \supset p')$ is true in *all* four cases. For *any* propositions p and q, the proposition $(p \supset q) \equiv (q' \supset p')$ is true. Such a proposition is known as a *tautology*. Tautologies are true in all possible cases. The purpose of propositional logic is to provide methods for determining which expressions are tautologies. Truth tables constitute one sure-fire method. Another method (more algebraic in character) is provided in Part III of this chapter.

Formulas – To approach our subject more rigorously, we need to define a *formula*. The letters p, q, r, with or without subscripts are called *propositional variables*; these are the simplest possible formulas, and they stand for unknown propositions (just as in algebra, the letters x, y, z, with or without subscripts, stand for unknown numbers). By a *formula* we mean any expression constructed according to the following rules:

(1) Each propositional variable is a formula.

(2) Given any formulas X and Y already constructed, the expressions

$X', (X \wedge Y), (X \vee Y), (X \supset Y)$ and $(X \equiv Y)$ are also formulas.

It is to be understood that no expression is a formula unless it is constructed according to rules (1) and (2) above.

When displaying a formula standing alone, we can dispense with outer parentheses without incurring any ambiguity. For example, when we say "the formula $p \supset q$", we mean "the formula $(p \supset q)$".

A formula in itself is neither true nor false, but only becomes true or false when we *interpret* the propositional variables as standing for definite propositions. We can, however say that a formula is *always true, never true* or *sometimes true and sometimes false*, if it is respectively true in all cases, true in no cases, true in some cases and false in some cases. For example, $p \vee p'$ is always true (it is a tautology); $p \wedge p'$ is always false, whereas $(p \vee q) \supset (p \cap q)$ is true in some cases (the cases when p and q are both true, or both false) and false in the other two cases. Formulas which are always false are called *contradictory* formulas, or more briefly *contradictions*. Formulas which are always true are called tautologies (as we have already indicated), and formulas which are true in some cases and false in others, are sometimes called *contingent*.

Some Tautologies – The truth table is a systematic method of verifying tautologies, but some tautologies are so obvious that they can be immediately perceived as such. Here are some examples:

(1) $((p \supset q) \land (q \supset r)) \supset (p \supset r)$

This says that if p implies q and q implies r, then p implies r. This is surely self-evident, though, of course, verifiable by a truth table. This tautology has a name–it is call the *syllogism*.

(2) $(p \land (p \supset q)) \supset q$

This says that if p and $p \supset q$ are both true, so is q. This is sometimes paraphrased: "Anything implied by a true proposition is true".

(3) $((p \supset q) \land q') \supset p'$

Thus, if p implies q and q is false, then p must be false. More briefly, "Any proposition implying a false proposition must be false". Thus, a true proposition can never imply a false one, and so we could write (3) in the equivalent form:

$(p \land q') \supset (p \supset q)'$

(4) $((p' \supset q) \land (p' \supset q')) \supset p$

This principle is known as *reductio ad absurdum*. To show that p is true, it suffices to show that p' implies some proposition q as well as its negation q'. No true proposition could imply both q and q', so if p' implies them both, then p' must be false, which means that p must be true. [Symbolic logic is, in the last analysis, merely a systematization of common sense].

(5) $((p \supset q) \land (p \supset r)) \supset (p \supset (q \land r))$

Of course, if p implies q and p implies r, then p must imply both q and r.

(6) $((p \lor q) \land ((p \supset r) \land (q \supset r))) \supset r$

This principle is known as *proof by cases*. Suppose $p \lor q$ is true. Suppose also that p implies r and q implies r. Then r must be true, regardless of whether it is p or q that is true (or both).

The reader with little experience in propositional logic should benefit from the following exercise.

Exercise 1 – State which of the following are tautologies, which are contradictions and which are contingent (sometimes true, sometimes false).

(a) $(p \supset q) \supset (q \supset p)$

(b) $(p \supset q) \supset (p' \supset q')$

(c) $(p \supset q) \supset (q' \supset p')$

(d) $(p \equiv q) \equiv (p' \equiv q')$

(e) $(p \supset p')$

(f) $(p \equiv p')$

(g) $(p \wedge q)' \equiv (p' \wedge q')$

(h) $(p \wedge q)' \equiv (p' \vee q')$

(i) $(p' \vee q') \supset (p \vee q)'$

(j) $(p \wedge q)' \supset (p' \vee q')$

(k) $(p' \vee q') \wedge (p \equiv (p \supset q))$

(l) $(p \equiv (p \wedge q)) \equiv (q \equiv (p \vee q))$

Answers –

(a) Contingent

(b) Contingent

(c) Tautology

(d) Tautology

(e) Contingent (see remarks below)

(f) Contradiction

(g) Contingent

(h) Tautology

(i) Contingent

(j) Tautology

(k) Contradiction

(l) Tautology (see remarks below)

Remarks – (1) Concerning (e), many beginners fall into the trap of thinking that (e) is a contradiction. They think that no proposition can imply its own negation. This is not so; if p is *false*, then p' is true, hence $p \supset p'$ is then true $(F \supset T = T)$. Thus, when p is *true*, then $(p \supset p')$ is false, but when p is false, then $p \supset p'$ is true. So $p \supset p'$ is true in one case and false in the other.

(2) Concerning (1), both $p \equiv (p \wedge q)$ and $q \equiv (p \vee q)$ have the same truth tables as $p \supset q$.

Logical Implication and Equivalence – A formula X is said to *logically imply* a formula Y if Y is true in all cases in which X is true, or what is the same thing, if $X \supset Y$ is a tautology. Formulas X and Y are said to be *logically equivalent* if they are true in exactly the same cases, or what is the same thing, if $X \equiv Y$ is a tautology, or again what is the same thing, if X and Y have the same truth table (in their last columns).

Finding a Formula, Given Its Truth Table – Suppose I tell you what the distribution of T's and F's is in the last column of the truth table, can you find a formula having that as its truth table? For example, suppose I consider a case of three variables p, q, and r, and I write down at random T's and F's in the fourth column thus:

p	q	r	?
T	T	T	F
T	T	F	F
T	F	T	T
T	F	F	F
F	T	T	F
F	T	F	T
F	F	T	F
F	F	F	T

The problem is to find a formula whose last column of its truth table is that under the question mark.

Do you think that cleverness and ingenuity are required? Well, it so happens that there is a ridiculously simple mechanical method that solves all problems of this type! Once you realize the method, then whatever distribution of T's and F's in the last column is given, you can instantly write down the required formula.

Problem 1 – What is the method?

Interdependence of the Logical Connectives – Suppose someone says he understands the meaning of ' (not) and $\wedge$ (and), but doesn't understand the meaning of $\vee$ (or). Could one explain it to him in terms of $\wedge$ and ' ? That is, can one construct a formula using just the logical connectives $\wedge$ and ' which is equivalent to the formula $p \vee q$?

Problem 2 – The answer to the above question is *yes*. Find such a formula.

Problem 3 - Also, if one starts with just ′ and ∨, one can define ∧–that is, there is a formula using just ′ and ∨ which is equivalent to $p \wedge q$. What formula would work?

Problem 4 – (a) How does one define ⊃ in terms of ′ and ∧ ? (b) How does one define ⊃ in terms of ′ and ∨ ?

Problem 5 – Starting with just ′ and ⊃, one can get (define) both ∧ and ∨. How?

Problem 6 – How does one get ≡ from ⊃ and ∧ ? Also, what is a direct way to get ≡ from ∧, ∨ and ′ ?

Sheffer's Discovery – In 1913, H. M. Sheffer pointed out that there is a *single* two-place logical connective from which one can attain ∧, ∨, ⊃ and ≡. Actually, there are *two* solutions (and only two), one of which is due to Sheffer.

Problem 7 – What are the two solutions?
  Further Interdependence Results

Problem 8 – We have seen how to get ∨ from ′ and ⊃ ($p \vee q$ is equivalent to $p' \supset q$), but it so happens that ∨ can be defined from ⊃ alone! How? [The solution is quite tricky!]

Problem 9 – It is also possible to define ∧ from ⊃ and ≡. How?

Problem 10 – One can also get ∧ from ≡ and ∨. How?

Problem 11 – One can also get ⊃ from ≡ and ∨. How?

Problem 12 – One can also get ⊃ from ≡ and ∧. How?

Problem 13 – One can also get ∨ from ≡ and ∧. How?

Remark – The astute reader might realize at this point that the last few problems are essentially the same as some of those in the earlier chapter, "Some Neighboring Gardens". The relation between propositional logic and Boolean flower gardens will be fully explained shortly.

Inequivalence – By $p \not\equiv q$ is meant $(p \equiv q)'$–p is *not* equivalent to q. It has the following truth table:

p	q	$p \not\equiv q$
T	T	F
T	F	T
F	T	T
F	F	F

Thus $p \not\equiv q$ is the opposite of $p \equiv q$, and $p \not\equiv q$ is true if and only if p and q have *different* truth values (one of them is true and the other is false). This is the same as the *exclusive* disjunction of p with q (one and *only* one of them is true). As we can read $p \equiv q$ as "p is equivalent to q", we can also read $p \not\equiv q$ as "p is *inequivalent* to q". Thus *inequivalent* is the same as *not equivalent*.

Formulas Involving T and F – It will now be convenient to extend our notion of "formula" by defining a *simple* formula as either a propositional variable, or the letter T or the letter F. Compound formulas are then constructed from simple formulas as before. Thus our new rules are:

(1) Propositional variables and T and F are formulas (called *simple* formulas).

(2) For any formulas X and Y, the expressions X', $(X \wedge Y)$, $(X \vee Y)$, $(X \supset Y)$ and $(X \equiv Y)$ and $X \not\equiv Y$ are again formulas.

It is to be understood that T stands for *truth* and F for *falsehood*, and so in any interpretation of a formula involving T and F, we must always interpret T to be some *true* proposition, and F to some *false* one. Also, in constructing a truth table for a formula involving T and/or F, it is understood that under the T heading, the entire column must consist of all T's, and under the F heading, the entire column must consist of F's. For example, here is a truth table for the tautology $((T \supset p) \wedge (q \supset F)) \supset (p \wedge q)'$.

p	q	T	F	$T \supset p$	$q \supset F$	$(T \supset p)\wedge$ $(q \supset F)$	$p \wedge q$	$(p \wedge q)'$	$((T \supset p)\wedge$ $(q \supset F)) \supset$ $(p \wedge q)'$
T	T	T	F	T	F	F	T	F	T
T	F	T	F	T	T	T	F	T	T
F	T	T	F	F	F	F	F	T	T
F	F	T	F	F	T	F	F	T	T

Here are some more tautologies involving T and/or F.

$(p \wedge T) \equiv p$

$(p \vee T) \equiv T$

$(T \vee T) \equiv T$

$(T \vee F) \equiv T$

$(F \vee F) \equiv F$

$(p \vee p') \equiv T$

$(p \wedge p') \equiv F$

$(T \supset p) \equiv p$

$(p \supset F) \equiv p'$

Problem 14 – Some treatments of propositional logic start with just $\supset$ and F, and then define from them the connectives $\wedge, \vee,',\supset$ and $\equiv$. How can this be done?

Some Special Tautologies – The following tautologies (all verifiable by truth tables or by common sense) will play a special role in Chapter 21.

T_1: $(p \wedge q) \equiv (q \wedge p)$

T_2: $(p \wedge (q \wedge r)) \equiv ((p \wedge q) \wedge r)$

T_3: $(p \wedge (q \vee r)) \equiv ((p \wedge q) \vee (p \wedge r))$

T_4: $(p \vee q)' \equiv (p' \wedge q')$

T_5: $p'' \equiv p$

T_6: $(p \wedge p') \equiv F$

T_7: $(p \wedge F) \equiv F$

T_8: $(p \wedge T) \equiv p$

T_9: $F' \equiv T$

II – Propositional Logic and Boolean Gardens

Now, we shall turn to the relation of Propositional Logic to Boolean gardens. Instead of propositions p, q, r ... of propositional logic, we dealt with *flowers* A, B, C ... in the magic garden of George B., and instead of the two truth values *truth* and *falsity* of propositional logic, we dealt with the two colors, *blue* and *red*, where now let us think of *blue* as corresponding to *true*, and *red* as corresponding to *false*. For any two flowers A and B, there was given the flower $A \mid B$ which is red an just those days when A and B are both blue–or what is the same thing, $A \mid B$ is blue when and only when A and B are not both blue–just as in propositional logic, $p \mid q$ is *true* if and only if p and q are not both *true*. And thus, our starting point in Chapter 13 (Condition B) was really with the Sheffer stroke operation!

The flower $\overline{A}$ is always of different color than A, just as in propositional logic, p' is of different truth value than p. Thus, the bar operation A (A bar) corresponds to *negation* in propositional logic.

Likewise, the cup operation $\cup$ for flower corresponds to the disjunction operation $\vee$ for propositions ($F \cup G$ is blue providing at least one of F, G is blue, just as $p \vee q$ is true, providing that at least one of p, q is true).

The operation $\supset$ for flowers corresponds to $\supset$ for propositional logic since $A \supset B$ is blue just when either A is red or B is blue, just as $p \supset q$ is true just in case either p is false or q is true. Also the flower $A \equiv B$ is blue just when A and B are of the same color, just as $p \equiv q$ is true just in case p and q have the same truth value (both true or both false). Thus, the operation $\supset$ on flowers corresponds to the operation $\supset$ on propositions, and $\equiv$, applied to flowers, corresponds to $\equiv$ applied to propositions.

The parallelism is now complete and yields a most important principle to which we now turn.

Given a formula X of propositional logic, if we reinterpret the propositional variables p, q, r, as names of flowers of the garden (instead of propositions, as in propositional logic) and we reinterpret the symbols $\wedge, \vee,', \supset, \equiv$ as the flower operations $\cap, \cup, -, \supset, \equiv$ respectively, then X designates some flower in the garden. [For example, if we interpret p as flower A and q as flower B, then under that interpretation, the formula $p \vee (q \wedge p')$ designates the flower $A \cup (B \cap \overline{A})$. Now, by the parallelism that we have seen above, it follows that if X is any tautology, then under every possible interpretation, formula X will name a flower which is always blue. Thus, for example, since $p \vee p'$ is a tautology, then for every flower A, the flower $A \cup \overline{A}$ is always blue.

As a consequence of this simple but vital principle, it follows that if X and Y are logically equivalent formulas, then under every interpretation of the proposition variables of X and Y, the formulas X and Y will name flowers that are *similar* (same color on all days); for suppose X and Y are logically equivalent, then $X \equiv Y$ is a tautology. Under any interpretation, X names some flower A and Y names some flower B, hence $X \equiv Y$ names $A \equiv B$, but since $X \equiv Y$ is a tautology, then the flower $A \equiv B$ named by $X \equiv Y$ is always blue, and hence A and B are always of the same color. Thus, for example, since the formula $p \wedge (q \vee r)$ is logically equivalent to $(p \wedge q) \vee (p \wedge r)$, it follows that for any flowers A, B and C of a Boolean garden, the flower $A \cap (B \cup C)$ is similar to the flower $(A \cap B) \cup (A \cap C)$.

We thus can use propositional logic to prove various relationships among flowers of a Boolean Garden.

We recall that by a *reduced* Boolean garden we mean one such that similar flowers are identical. So let us now note for future reference that in a *reduced* Boolean garden, if X and Y are logically equivalent formulas, then under any interpretation of X and Y, the formulas X and Y designate the very same flower!

Exercise 2 – Suppose that under a certain interpretation, Formula X designates flower A and Formula Y designates flower B. Prove that if X logically implies Y ($X \supset Y$ is a tautology), then flower B dominates flower A (B is blue on all days on which A is blue).

Exercise 3 – Show how truth tables can be used to solve all the exercises of Chapter 13.

III – An Algebraic Approach to Propositional Logic

So far, we have approached propositional logic from the viewpoint of truth tables. We now consider a quite different approach that will be important in Chapter 21. This approach is interesting in its own right and brings to light both some similarities and differences between laws of propositional logic and well-known laws of arithmetic.

We shall use the word "number" to mean either zero or one of the positive whole numbers 1, 2, 3, 4, ... [These are the so called *natural* numbers][1]. For two such numbers x and y, by xy is meant the *product* of x and y–that is, the result of multiplying x by y.

Two numbers are said to be of the same *parity* if they are either both even or both odd. We know the following facts:

(1) If x and y are both odd, so is xy.

(2) If x is odd and y is even, then xy is even.

(3) If x is even and y is odd, then xy is even.

(4) If x is even and y is even, then xy is even.

These facts can be summarized by the following even-odd table (0 stands for *odd*, and E for *even*).

x	y	xy
0	0	0
0	E	E
E	0	E
E	E	E

Eureka! This is exactly the truth table for *conjunction*, if we replace 0 by T, E by F and xy by $x \wedge y$ (and if we think of x and y as *propositions* instead of numbers). Actually this is not all that surprising, since in fact xy is odd if and only if x is odd *and* y is odd.

Thus, we now let oddness correspond to *truth* and evenness to *falsehood*, and then multiplication corresponds to *conjunction*. We are now going to interpret the propositional variables p, q, r, etc. as *numbers* instead of propositions.

---

[1]Someone once asked me to give an example of an *unnatural* number.

We regard zero as an even number, and so it is one of the numbers associated with falsehood, and 1, being odd, is associated with truth. We let F be 0 (zero) and T be 1.

What should we choose to correspond to negation? That is, what operation on numbers should we take that will convert an even number to an odd one, and an odd number to an even one? The obviously simple choice is to add one. And so we will take p' to be p+1. Next what about disjunction? Well, we know that $p \lor q$ is logically equivalent to $(p' \land q')'$, and so we could interpret $p \lor q$ as $((p+1)(q+1))+1$. However, this can be simplified to $(pq+p+q+1)+1$, which is $pq+p+q+(1+1)$, which is $pq+p+q+2$. But 2 is even, so $pq+p+q+2$ has the same parity as $pq+p+q$ and so we shall more simply take $p \lor q$ to be $pq+p+q$.

What is $p+q$ itself? Well, the sum of two numbers is odd if and only if one of them is even and the other odd (two odds added together is even, and of course, two evens added together is even). And so we have the following table:

p	q	p+q
0	0	E
0	E	0
E	0	0
E	E	E

If we replace 0 by T and E by F, we have the truth table for $\not\equiv$ (inequivalence, or exclusive disjunction). It then follows that $p \equiv q$, which is the opposite, can be taken to be $p+q+1$. And so we henceforth interpret $p \equiv q$ as $p+q+1$.

What about $\supset$? Well, $p \supset q$ is equivalent to $p' \lor q$, which **now is** $(p+1)q + (p+1) + q$, which simplifies to $pq + q + p + 1 + q$. However, $q+q$ is even, hence we can delete it and thus get $pq + p + 1$.

We thus now interpret the propositional connectives and T and F as follows:

(1) $p' = p+1$

(2) $p \land q = pq$

(3) $p \lor q = pq + p + q$

(4) $p \supset q = pq + p + 1$

(5) $p \equiv q = p + q + 1$

(6) $p \not\equiv q = p + q$

(7) $T = 1$

(8) $F = 0$

We can now reduce problems in propositional logic to problems in arithmetic—problems of whether certain combinations of numbers are always odd. We can use the usual laws of arithmetic and, in addition, two laws *not* present in ordinary arithmetic.

(1) x^2 (which is xx) has the same parity as x, hence we can always replace x^2 by x. Thus exponents no longer matter!

(2) x+x is always even, hence can be replace by 0.

We then have an alternative test to that of the truth table for determining whether or not a given formula is a tautology. We go through the following steps (where now, the propositional variable p, q, r, ... represent arbitrary *numbers*, instead of arbitrary propositions. For any formulas A, B that are parts of the formula that you are testing),

(1) Replace A' by $A + 1$.

(2) Replace $A \wedge B$ by AB.

(3) Replace $A \vee B$ by $AB + A + B$.

(4) Replace $A \supset B$ by $AB + A + 1$.

(5) Replace $A \equiv B$ by $A + B + 1$.

(6) Replace $A \not\equiv B$ by $A + B$.

(7) Replace T by 1.

(8) Replace F by 0.

Then use the usual laws of arithmetic and, in addition, AA can be replaced by A and A+A by 0. If you then can reduce the whole expression to 1, then you have a tautology, otherwise not.

Let's consider some examples, First, let's try the formula $(p \vee q)' \equiv (p' \wedge q')$. This becomes $((pq+p+q)+1)+((p+1)(q+1))+1$, which is $pq+p+q+1+(pq+p+q+1)+1$, which is $pq+pq+p+p+q+q+1+1+1$, which reduces to $0+0+0+0+1$, which is 1. Thus, $(p \vee q)' \equiv (p' \wedge q')$ reduces to 1, which is odd, and we have a tautology.

Actually, we can save much labor by a shortcut: To prove $X \equiv Y$, instead of trying to reduce the whole formula $X \equiv Y$ to 1, reduce X and Y separately and see if they reduce to the same thing. Well, in the above example $(p \vee q)' = (pq+p+q)+1$, whereas $p' \wedge q' = (p+1)(q+1) = pq+p+q+1$. Thus, $(p \vee q)'$ and $p' \vee q'$ both reduce to $pq+p+q+1$, and so they are equivalent.

What about a simple one such as $p \supset p$? Well, this is $pp+p+1$, but pp can be replaced by p and so $pp+p+1$ reduces to $p+p+1$, which reduces to 1 (since $p+p$ reduces to zero).

Let's now try showing that $p \wedge (q \vee r)$ is logically equivalent to $(p \wedge q) \vee (p \wedge r)$. Well, $p \wedge (q \vee r) = p(qr+q+r) = pqr + pq + pr)$, whereas $(p \wedge q) \vee (p \wedge r) = (pq)(pr) + pq + pr = p^2qr + pq + pr$, which reduces to $pqr + pq + pr$, and so the logical equivalent holds.

Some More Special Tautologies – The following group of tautologies will also have a special significance in Chapter 21.

$S_1 : (p \wedge q) \equiv (q \wedge p)$

$S_2 : (p \wedge (q \wedge r)) \equiv ((p \wedge q) \wedge r)$

$S_3 : (p \not\equiv q) \equiv (q \not\equiv p)$

$S_4 : (p \not\equiv (q \not\equiv r)) \equiv ((p \not\equiv q) \not\equiv r)$

$S_5 : (p \wedge (q \not\equiv r)) \equiv ((p \wedge q) \not\equiv (p \wedge r))$

$S_6 : (p \not\equiv F) \equiv p$

$S_7 : (p \wedge T) \equiv p$

$S_8 : (p \wedge p) \equiv p$

$S_9 : (p \not\equiv p) \equiv F$

$S_{10} : p' \equiv (p \not\equiv T)$

$S_{11} : (p \vee q \equiv ((p \wedge q) \not\equiv (p \not\equiv q)))$

These tautologies can be verified by truth tables, but much more swiftly by the algebraic method that we have described. [For example, S_2 simply boils down to showing that $p(qr)$ has the same parity as $(pq)r$, which is immediate, since they are the same number].

Exercise 4 – By the algebraic method, verify R_1–R_{11}.

Exercise 5 – By the algebraic method, verify the tautologies T_1–T_9 at the end of Part I.

SOLUTIONS

1 – This particular case will illustrate the general method perfectly!

In this case, the formula is to come out T in the third, sixth and eighth rows. Well, the third row is the case when p is true, q is false and r is true–in other words, when $(p \wedge q' \wedge r)$ is true. The sixth case is the case when $(p' \wedge q \wedge r')$ is true, and in the eighth case $(p' \wedge q' \wedge r')$ is true. Thus, the formula is to be true when and only when *at least one* of those three cases holds, and so a solution is simply

$$(p \wedge q' \wedge r) \vee (p' \wedge q \wedge r') \vee (p' \wedge q' \wedge r')$$

2 – To say that at least one of p, q is true is to say that it is not the case that both are false–in other words, that p' and q' are not both true. Thus, $p \vee q$ is equivalent to $(p' \wedge q')'$.

3 – To say that $p \wedge q$ is true is to say that neither p nor q is false, and so $p \wedge q$ is equivalent to $(p' \vee q')'$ (as can be easily verified by a truth table.).

4 – $p \supset q$ has the same truth table as $(p \wedge q')'$. It also has the same truth table as $p' \vee q$.

5 – Take $p \wedge q$ to be $(p \supset q')'$. [Thus $\wedge$ is definable from $\supset$ and $'$]. To define $\vee$ from $\supset$ and $'$, simply take $p \supset q$ to be $p' \vee q$.

6 – $p \equiv q$ is equivalent to $(p \supset q) \wedge (q \supset p)$. Also, $p \equiv q$ is equivalent to $(p \wedge q) \vee (p' \wedge q')$.

7 – Sheffer defined $p \mid q$ to mean that p and q are not both true. It has the following truth table:

p	q	$p \mid q$
T	T	F
T	F	T
F	T	T
F	F	T

Thus $p \mid q$ is the very opposite of $p \wedge q$. Now, starting with Sheffer's "stroke" operation $\mid$, we can take p' to be $p \mid p$, then take $p \wedge q$ to be $(p \mid q)'$. Once we have $'$ and $\wedge$, we can then get $\vee, \supset$ and $\equiv$, as already indicated.

The other well-known solution (and it happens to be the *only* other one) is to take $p \downarrow q$ to mean that p and q are both false. The operation $\downarrow$ is aptly called *joint denial* and is the very opposite of disjunction, and has the following truth table:

p	q	$p \downarrow q$
T	T	F
T	F	F
F	T	F
F	F	T

Starting with $\downarrow$, we can take p' to be $p \downarrow p$, and then $p \vee q$ to be $(p \downarrow q)'$. Once we have $\vee$ and $'$, we can get $\wedge, \supset$ and $\equiv$, as already indicated.

8 – $p \vee q$ is equivalent to $(p \supset q) \supset q$. [I don't know who discovered this!]

9 – $p \wedge q$ is equivalent to $p \equiv (p \supset q)$. [I discovered this independently. I do not know if it has or has not been previously discovered by someone else.]

10 – $p \wedge q$ is equivalent to $(p \vee q) \equiv (p \equiv q)$

11 – $p \supset q$ is equivalent to $q \equiv (p \vee q)$

12 – $p \supset q$ is equivalent to $p \equiv (p \wedge q)$

13 – $p \vee q$ is equivalent to $(p \wedge q) \equiv (p \equiv q)$

14 – $\sim p$ is equivalent to $p \supset f$. Once one has $\sim$ and $\supset$, one can get $\wedge, \vee$ and $\equiv$.

Chapter 18

The Boolean Theory of Sets

I – The Basic Operations on Sets

The notion of *set* is so basic, that it is hopeless to try to define it in terms of anything more basic. A set is any group or bunch of objects whatsoever. For any set S and object or element x, we say that x *belongs* to S, or is a *member* of S, or that S *contains* x if x is one of the elements of S. We symbolize the statement "x is a member of S" by "$x \in S$". Thus "$\in$" is read: "is a member of". We write "$x \notin S$" to mean that x is *not* a member of S. Thus $\notin$ abbreviates "is not a member of". If x is a member of S, then we also say that x is *inside* S, and if x is not a member of S, we also say that x is *outside* S.

A set A is said to be a *subset* of a set B–in symbols, $A \subseteq B$–if every element of A is also an element of B. For example, if H is the set of all humans and W is the set of all women, then $W \subseteq H$ is true, but $H \subseteq W$ is false (not every human is a woman). We recall that a set is called *empty* if it has no members at all. There is only one empty set and its symbol (or at least one of its symbols) is $\emptyset$. A set is called *non-empty* if it has at least one member.

For sets A and B, the only way that A can fail to be a subset of B is that A contains at least one element not in B. Now, the empty set $\emptyset$ cannot contain any element not in B, since $\emptyset$ doesn't contain any elements at all; hence, $\emptyset$ cannot fail to be a subset of B and so $\emptyset$ *is* a subset of B. Thus, we have the curious but vital fact that the empty set is a subset of *every* set! For every set B, it is true that $\emptyset \subseteq B$.

Another way of looking at the matter is this: To say that $A \subseteq B$ is to say that for every element x, the implication $x \in A \supset x \in B$ holds. Now, $x \in \phi$ is false, hence by propositional logic, $x \in \phi \supset x \in B$ is true (remember that for any false proposition F and any propositions q, the implication $F \supset q$ holds), and so $\phi \subseteq B$.

For any element x, by $\{x\}$ is meant the set whose only element is x. [This is not to be confused with x itself. For example, ϕ is different from $\{\phi\}$, since $\{\phi\}$ contains an element (namely, ϕ), whereas ϕ doesn't contain any elements]. By $\{x, y\}$ is meant the set whose only elements are x and y. By $\{x, y, z\}$ is meant the set whose only elements are x, y and z, and similarly for finite sets of four or more elements. [Incidentally, it might be amusing and instructive to note that another symbol for the empty set is { }].

As with the special case of basic flowers considered in Chapter 15, for any two sets A and B, by the *intersection* of A and B–symbolized $A \cap B$–is meant the set of all things that are in both A and B. For exemple, if A is the set of all white things and B is the set of all birds, then $A \cap B$ is the set of all things that are both white and birds–in other words, the set of all white birds. And by the *union* of A and B–symbolized $A \cup B$–is meant the set of all things that are either in A or in B (or both). For exemple, if A is the set of all even numbers and B is the set of all numbers divisible by 5, then $A \cup B$ consists of all even numbers together with all multiples of 5. [Thus the first twelve members of this set are 0, 2, 4, 5, 6, 8, 10, 12, 14, 15, 16, 18. Note that 10 is both in A and in B. What is the intersection of $A \cap B$ of A and B?

Answer: All numbers that are both even and multiples of 5–in other words, all multiples of 10].

We now consider a set I fixed for the discussion, which we call the *universe of discourse*. What the set I is, varies from one application to another–for example, in propositional logic, I will be the set of all formulas, whereas in Boolean gardens, I can be the set of all the flowers (or perhaps just the set of basic flowers, as in Chapter 15, or in number theory, I is the set of all numbers under discussion, whereas in plane geometry, I might be the set of all points in the plane. For now, I is a completely arbitrary set, since we are building a *general* theory of sets, and we will be studying the collection of all *subsets* of I (which is a Boolean algebra, as will be defined in the next chapter). The largest subset of I is I itself, and the smallest subset of I is the empty set $\emptyset$. If I itself is empty, then $\emptyset$ is the only subset of I. If I contains just one element, then there are just 2 subsets of I–namely, $\emptyset$ and I itself. As we explained in Chapter 15, if I is a finite set with n elements, then the number of subsets of I is 2^n. From now on, we will assume that I is non-empty.

For any subset A of our universe *I*, by its *complement* A' (relative to *I* understood)–also written $\overline{A}$–is meant the set of all elements of I that are *not* in A. So, for example, if I is the set of natural numbers and E is the set of even numbers, then E' is the set of odd numbers. We note that E'' is the complement of E' – the complement of the set of odd numbers–which is the set E again. Thus E''=E–and in general, for any subset A of our universe $I, A'' = A$. We also note that $\phi' = I$ and $I' = \phi$. (Why?)

The operations of intersection, union and complementation can be graphically illustrated by what are called *Venn diagrams*: In these, *I* is represented by the set of all points in the interior of a square and A and B are subsets of *I*.

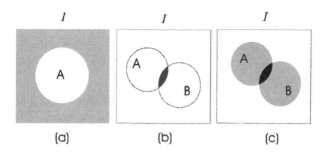

(a)   (b)   (c)

In (a), the shaded portion is the complement A' of A. In (b), the shaded portion is the intersection $A \cap B$, and in (c), the shaded portion is the union $A \cup B$.

Problem 1 – Which of the following statements are correct?

(a) $A \subseteq B$ if and only if $A \cap B = B$

(b) $A \subseteq B$ if and only if $A \cap B = A$

(c) $A \subseteq B$ if and only if $A \cup B = A$

(d) $A \subseteq B$ if and only if $A \cup B = B$

Problem 2 – Which of the following statements are true for all sets A and B?

(a) $A \subseteq B$ implies $B \subseteq A$.

(b) $A \subseteq B$ implies B is not a subset of A.

(c) $A \subseteq B$ implies $B' \subseteq A$.

(d) $A \subseteq B$ implies $B \subseteq A'$.

(e) $A \subseteq B$ implies $B' \subseteq A'$.

(f) $A \subseteq B$ if and only if $B' \subseteq A'$.

(g) $A \subseteq B'$ if and only if $B \subseteq A'$.

Problem 3 – Is the following statement necessarily true?
$A \subseteq B$ if and only if $A' \cup B = I$.

II – Relation to Boolean Gardens

Let us note the following about these operations on sets and the corresponding operations on flowers of a Boolean garden: For any flower F, let D(F) be the set of days on which F is blue. Then for any flowers F and G:

(1) $D(F \cap G) = D(F) \cap D(G)$

(2) $D(F \cup G) = D(F) \cup D(G)$

(3) $D(\overline{F}) = (D(F))'$

(1) says that the set of days on which $F \cap G$ is blue is the intersection of the set of days on which F is blue with the set of days on which G is blue–in other words, $F \cap G$ is blue on just those days in which F and G are both blue.

(2) says the same thing, with *union* in place of intersection.

(3) says that the set of days on which $\overline{F}$ is blue is the *complement* of the set of days on which F is blue–*i.e.* the set of days on which F is *not* blue.

We recall that in Chapter 15, we said that F *dominates* G if F is blue on all days that G is blue. We write $G \leq F$ to mean that F dominates G.

Exercise 1 – Prove that $G \leq F$ iff $D(G) \subseteq D(F)$

Let us call a subset A of I *atomic* (or *basic*) if $A \neq \emptyset$ and if the only subsets of A are A and $\emptyset$.

Problem 4 – Which, if any, of the following three statements are necessarily true?

(1) If A is atomic, then A contains exactly one element.

(2) If A contains exactly one element, then A is atomic.

(3) If A is non-empty, then some subset of A is atomic.

Set Equations – We use letter A, B, C, with or without subscripts, as standing for arbitrary sets (just as in propositional logic we use p, q, r, with or without subscripts, as standing for arbitrary propositions). We call these letters A, B, C (with or without subscripts) *set variables*. By a *term* we mean any expression constructed according to the following rules:

(1) Each variable standing alone is a term.

(2) For any terms t_1 and t_2, the expressions $(t_1 \cap t_2), (t_1 \cup t_2)$ and t_1' are again terms.

Examples of terms are $A \cup (B \cap C')'$ or $A' \cup (B \cap A'')'$. We sometimes delete some parentheses when no ambiguity can result. [For example, we wrote $A \cup (B \cap C')'$ for $(A \cup (B \cap C')')$].

By a (Boolean) set-equation (more briefly, an *equation*) is meant an expression of the form $t_1 = t_2$, where t_1 and t_2 are terms. Here are some examples of set-equations (some of which are valid and some are not).

(1) $A \cup B = A \cap B$

(2) $A' = B$

(3) $A \cup B = B \cup A$

(4) $(A \cup B)' = A' \cup B'$

(5) $(A \cup B)' = A' \cap B'$

(6) $A \cap (B \cup C) = A \cup (B \cap C)$

(7) $A \cap (B \cup C) = (A \cap B) \cup (A \cap C)$

An equation is called *valid* if it is true for *all* values of the set variables. For example (3) is valid, since for any sets A and B, $A \cup B = B \cup A$. Equation (1) is certainly *not valid*. (It is *not* true that for *all* sets A and B, the sets $A \cup B$ and $A \cap B$ are the same; they are only when A and B happen to be the same set). Equation (2) is obviously not valid. Equation (4) isn't valid, but (5) is (as we will see). (6) is not valid, but (7) is.

In this chapter we will give two different systematic methods of establishing validity of set equations.

III – Relation to Propositional Logic

The operations of intersection, union and complementation on sets, correspond to the logical operations of conjunction, disjunction and negation respectively, by virtue of the following facts (where A, B are any subsets of I, and x is any element of I).

$F_1 : x \in (A \cap B)$ if and only if $(x \in A) \wedge (x \in B)$

$F_2 : x \in (A \cup B)$ if and only if $(x \in A) \vee (x \in B)$

$F_3 : x \in A'$ if and only if $x \notin A$ (*i.e.* iff $(x \in A)'$)

By virtue of the above three facts, we can apply propositional logic to the algebra of sets, which I will first illustrate by an example.

Suppose we wish to demonstrate that for any sets A and B, the identity $(A \cap B)' = A' \cup B'$ holds (which is one of De Morgan's laws). We recall that two sets are identical iff they contain the same elements, and so the equation $(A \cap B)' = A' \cup B'$ says nothing more nor less than that for element x of I,

$$(*) \quad x \in (A \cap B)' \equiv x \in (A' \cup B')$$

Now: (1) $x \in (A \cap B)'$ iff $(x \in A \wedge x \in B)'$ (because $x \in (A \cap B)'$ iff $(x \in A \cap B)'$, but $x \in A \cap B$ iff $(x \in A \wedge x \in B)$.

(2) $x \in A' \cup B'$ iff $(x \in A)' \vee (x \in B)'$ (because $x \in A' \cup B'$ iff $x \in A' \vee x \in B'$, but $x \in A'$ iff $(x \in A)'$, and $x \in B'$ iff $(x \in B)'$.

So by virtue of (1) and (2), the statement (∗) can be rewritten:

$$(\ast\ast) \quad (x \in A \wedge x \in B)' \equiv ((x \in A)' \vee (x \in B)')$$

But (∗∗) is a tautology! [It is the special case of the tautology $(p \wedge q)' \equiv (p' \vee q')$, where p is the proposition $x \in A$ and q is the proposition $x \in B$]. And so (∗∗) is true for all x and so (∗), which means that the set-equation $(A \cap B)' = A' \cup B'$ is valid.

Now, we don't have to go through all this work each time we check the validity of a set-equation! All we have to do to the equation is to replace $\cap$ by $\wedge$, $\cup$ by $\vee$ and $=$ by $\equiv$ and (to be pedantic) set variables by propositional variables, and if we then get a tautology, the equation is valid. For example, in the equation $(A \cap B)' = (A' \cup B')$, if we transform the equation in the above manner, we get the tautology $(p \wedge q)' \equiv (p' \vee q')$.

Other examples:

(1) $A'' = A$, since $p'' \equiv p$ is a tautology

(2) $(A \cup B)' = (A' \cap B')$, since $(p \wedge q)' \equiv (p' \wedge q')$ is a tautology.

[This law, by the way, is De Morgan's other law].

Going in the other direction, given any tautology of the form $X \equiv Y$, when X and Y are formulas using just the logical connectives $\wedge, \vee$ and $'$, if we replace $\wedge$ by $\cap$, $\vee$ by $\cup$, $\equiv$ by $=$ and (to be pedantic) propositional variables by set variables, we get a valid set-equation. For example:

(3) Since $(p \wedge (q \vee r)) \equiv ((p \wedge q) \vee (p \wedge r))$ is a tautology, then the equation $A \cap (B \cup C) = (A \cap B) \cup (A \cap C)$ is valid.

Extension of terms to include I and $\emptyset$ – let us now extend our notion of *term* by starting not only with set variables, but also with the constants I and $\emptyset$. Of course, $x \in I$ is always true (for all x in I) and $x \in \emptyset$ is always false, and so I and $\emptyset$ correspond to the propositional constants T and F respectively. And so, in a set-equation involving I and/or $\emptyset$, we replace I by T and $\emptyset$ by F (as well as $\cap$ by $\wedge$, $\cup$ by $\vee$ and $=$ by $\equiv$) and see if we get a tautology. For example, the equation $\emptyset' = I$ is valid, since $F' \equiv T$ is a tautology. Or again, since $(p \wedge T) \equiv p$ is a tautology, then the equation $A \cap I = A$ is valid (which in itself is obvious). Also, $A \cap A' = 0$ is valid, since $(p \wedge p') \equiv F$ is a tautology.

By virtue of the special tautologies $T_1 - T_9$ of Chapter 17 (end of Section 1), we at once have the following valid set equations:

(1) $A \cap B = B \cap A$

(2) $A \cap (B \cap C) = (A \cap B) \cap C$

(3) $A \cap (B \cup C) = (A \cap B) \cup (A \cap C)$

(4) $(A \cup B)^1 = A^1 \cap B^1$

(5) $A'' = A$

(6) $A \cap A' = \emptyset$

(7) $A \cap \emptyset = \emptyset$

(8) $A \cap I = A$

(9) $\emptyset' = I$

These 9 equations are of interest in that *all* valid Boolean set equations (of which there are infinitely many) are derivable from just these nine! [This result is a special case of a "grand" result proved in Chapter 21.]

IV – Indexing

There is another method of verifying Boolean equations of sets, which is quite neat and is often used these days. [It is an improvement an the older method of Venn diagrams.]

As a simple starter let A and B be subsets of I.

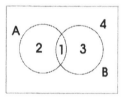

We see that I is divided up into the four sets of $A \cap B, A \cap B', A' \cap B$ and $A' \cap B'$, which we have labeled (indexed) by 1, 2, 3 and 4 respectively. Every element x of I belongs to one and only one of these four regions. Let us call these regions the *basic* regions. [They correspond to the basic flowers of Chapter 15.] Any combination of A and B, describable by use of $\cup, \cap$ and $'$, is the union of one and only one set of these basic regions (just as a basic flower is the union of one and only one set of basic flowers), and so there are 16 possible combinations altogether (2^4).

Now, let us identify any such combination with its set of indices. So for example,

A = (1, 2)

B = (1, 3)

Then $A \cup B = (1, 2, 3)$ and $A \cap B = (1)$ (since (1) is the only region common to A and B). Also $A' = (3, 4)$, and since there is nothing in common between $(1, 2)$ and $(3, 4)$, then $A \cap A' = \{\ \}$ (the empty set). Also, $A \cup A' = (1, 2) \cup (3, 4) = (1, 2, 3, 4)$ thus $A \cup A' = I$.

Now, suppose we wish to verify the De Morgan law, $(A \cup B)' = A' \cap B'$ by this method. The idea is to find first the set of indices of $(A \cup B)'$ and then the set of indices of $A' \cap B'$ and see if the two sets are the same.

$$A \cup B = (1, 2, 3) \qquad A' = (3, 4) \text{ and } B' = (2, 4)$$
$$\text{Hence } (A \cup B)' = (4) \qquad \text{Hence } A' \cap B' = (4)$$

Thus, (4) is the set of indices of both $(A \cup B)'$ and $A' \cap B'$, hence $(A \cup B)' = A' \cap B'$.

Let's now try an equation with three sets A, B and C. They divide I into eight basic regions thus:

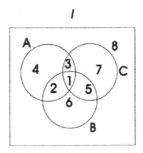

Thus

A = (1, 2, 3, 4)

B = (1, 2, 5, 6)

C = (1, 3, 5, 7)

Suppose we wish to show that $A \cup (B \cap A) = (A \cup B) \cap (A \cup C)$. Again we reduce each side of the equality sign to its set of indices and see if the two sets are the same.

$$B \cap C = (1, 5) \qquad A \cup B = (1, 2, 3, 4, 5, 6)$$
$$A \cup (B \cap C) = (1, 2, 3, 4, 5) \qquad A \cup C = (1, 2, 3, 4, 5, 7)$$
$$(A \cup B) \cap A \cup C) = (1, 2, 3, 4, 5)$$

Thus, both sides reduce to $(1, 2, 3, 4, 5)$ and we have thus won the case.

Let's try the equation $A \cap (B \cup C) = (A \cap B) \cup (A \cap C)$

$$B \cup C = (1, 2, 3, 5, 6, 7) \qquad A \cap B = (1, 2)$$
$$A \cap (B \cup C) = (1, 2, 3) \qquad A \cap B = (1, 3)$$
$$(A \cap B) \cup (A \cap C) = (1, 2, 3)$$

Thus, the equation reduces to $(1, 2, 3) = (1, 2, 3)$.

What do we do if we have more than three unknowns–say A, B, C and D? Well, we can no longer draw circles, but still the four sets divide I into 16 basic regions, and we can number them in such a way that

A = (1, 2, 3, 4, 5, 6, 7, 8)

B = (1, 2, 3, 4, 9, 10, 11, 12)

C = (1, 2, 5, 6, 9, 10, 13, 14)

D = (1, 3, 5, 7, 9, 11, 13, 15)

Then, we can operate accordingly with these sets of indices.

For five unknowns A, B, C, D, E we have 32 basic regions. In general, for any n equal or greater than 2, the sets $A_1, A_2, \ldots, A_n$ divide I into 2^n basic regions, and we can assign an index to each of these regions, say by taking A to be the first half of the integers $1, 2, \ldots, 2^n$, then taking A_2 to be every other quarter (starting with the first), A_3 every other eighth, and so forth. For example, for $n = 5$, we take $A_1 = (1 - 16)$;

$A_2 = (1 - 8, 17 - 25); A_3 = (1 - 4, 9 - 12, 17 - 21, 26 - 29)$;

$A_4 = (1, 2, 5, 6, 9, 11, 13, 14, 17, 19, 21, 23, 25, 27, 29, 31); A_5 = $ (All odd numbers from 1 to 31).

Which is more efficient, the tautology method or the indexing method? Well, this depends on the particular equation. Sometimes the indexing method works more quickly, and sometimes the tautology method.

Other Boolean Operations on Sets

(1) By $A \supset B$ is meant $A' \cup B$.

(2) By $A \equiv B$ is meant $(A \supset B) \cap (B \supset A)$.

(3) By $A - B$ is meant $A \cap B'$.

(4) By $A + B$ is meant $(A - B) \cup (B - A)$. [$A + B$ is sometimes called the *symmetric difference* of A and B. Note that $A + B = (A \equiv B)'$].

Exercise 2 – Prove the following facts below (where A and B are any subsets of I and x is any element of I).

$F_4:\ x \in A \supset B$ iff $(x \in A \supset x \in B)$

$F_5:\ x \in A \equiv B$ iff $(x \in A \equiv x \in B)$

$F_6:\ x \in A - B$ iff $(x \in A \cap (x \in B)')$

$F_7:\ x \in A + B$ iff $(x \in A \equiv B)'$

* * *

What we have done in this chapter–the Boolean algebra of sets–is but a special case of the theory of Boolean algebras in general, which we turn to in the next chapter.

SOLUTIONS

1 – (a) and (c) could be true for *some* sets A and B, but they are not generally true. On the other hand, (b) and (d) hold for *all* sets A and B, as we will now prove.

(b) Obviously $A \cap B \subseteq B$ (every element of $A \cap B$ is also in B, since it is in both A and B). Therefore, if $A \cap B$ is the same set as A, then $A \subseteq B$. Thus, $A \cap B = A$ implies $A \subseteq B$.

Conversely, suppose $A \subseteq B$. Then every element of A is also in B, hence is in $A \cap B$. Therefore, $A \subseteq A \cap B$. Also, of course, $A \cap B \subseteq A$, and so each of the sets $A \cap B$ and A is a subset of the other, and so $A \cap B = A$. This proves (b).

(d) If $A \cup B = B$, then the statement that $A \subseteq B$ reduces to $A \subseteq A \cup B$, which is obviously true. Thus, $A \cup B = B$ implies $A \subseteq B$. Conversely, suppose $A \subseteq B$. Then every element of $A \cup B$ is either in A or in B, but if it is in A, it is also in B (since $A \subseteq B$), and so every element of $A \cup B$ is in B, and therefore $A \cup B \subseteq B$. Also, of course, $B \subseteq A \cup B$, and so $A \cup B = B$. Thus, $A \subseteq B$ implies $A \cup B = B$.

2 – It is (e), (f) and (g) that are true for all sets A and B. Here are the proofs:

(e) Suppose $A \subseteq B$. Take any element x in B'. Then x is not in B. Hence x is not in A (because if it were in A, it would also be in B (since $A \subseteq B$), which it isn't). Since x is not in A, then $x \in A'$, and so every element of B' is in A', which means that $B' \subseteq A'$. Thus, $A \subseteq B$ implies $B' \subseteq A'$.

(f) First of all, it should be obvious (as mentioned earlier) that for any set $A, A'' = A$ (the complement of the complement of A is A again, because for any proposition p, the proposition p'' is equivalent to p, so take for p the proposition $x \in A$).

Now, suppose that $B' \subseteq A'$. Then $A'' \subseteq B''$ (by (e)), hence $A \subseteq B$. Thus, $B' \subseteq A'$ implies $A \subseteq B$. Also by (e), $A \subseteq B$ implies $B' \subseteq A'$. Hence, $A \subseteq B$ if and only if $B' \subseteq A'$.

(g) By (f), $A \subseteq B'$ if and only if $B'' \subseteq A'$. But $B'' = B$, and so $A \subseteq B'$ iff $B \subseteq A'$.

3 – Yes, it is true. Here is the proof:

Suppose $A \subseteq B$. To show that $A' \cup B = I$, it suffices to show that $I \subseteq A' \cup B$ (since obviously $A' \cup B \subseteq I$–we are dealing only with subsets of I). In other words, we must show that every element x of I is in $A' \cup B$. Well, if x is in A', then of course, x is in $A' \cup B$ (since $A' \subseteq A' \cup B$). On the other hand, if x is not in A', then x is in A, hence again x is in B (since $A \subseteq B$). So whether x is in A' or not, x is in $A' \cup B$. Thus, $I \subseteq A' \cup B$.

Conversely, suppose $A' \cup B = I$. Then every element x of I is in $A' \cup B$, hence $x \in A'$ or $x \in B$. Now, suppose x is in A. Then $x \in A'$ is not true, hence x must be in B (since $x \in A'$ or $x \in B$). Thus, every element x of A is also in B, so $A \subseteq B$.

4 – All three statements are true:

(1) Suppose A is atomic. Then $A \neq \emptyset$, hence A contains at least one element x. Then $\{x\}$ (the set whose only element is x) is a subset of A. If A contained any more elements, then A would be unequal to $\{x\}$, hence A would have the subset $\{x\}$, which is neither A nor $\emptyset$, contrary to the fact that A is atomic. Therefore, $A = \{x\}$.

(2) It is obvious that if A contains only one element, then $A = \{x\}$, where x is the sole element of A, and the only subsets of $\{x\}$ are $\{x\}$ itself and $\emptyset$, and so $\{x\}$ is atomic.

We note from (1) and (2) that an atomic set is nothing more nor less than a set with exactly one element.

(3) If A is non-empty, then it contains some element x, and so $\{x\}$ is an atomic subset of A.

Chapter 19

Boolean Algebras in General

It has been wisely remarked that the existence of analogies between the central features of various theories implies the existence of a general theory which underlies the particular theories and unifies them with respect to those central features[1].

Well, we are about to consider a general abstract theory that unites the theories of Boolean gardens, Propositional Logic and the Boolean algebra of sets.

I – Boolean Algebra

We now consider a set $\cup$ of at least two elements, together with an operation $\wedge$ that assign to each element x of $\cup$ and each element y of $\cup$ a unique element $x \wedge y$ of $\cup$, called the *meet* of x and y, and another operation $\vee$ that assigns to each x and y an element $x \vee y$ of $\cup$, called the *join* of x and y, and an operation that assigns to each x in $\cup$ an element x' of $\cup$, which might be called the *involute* of x. We let B be the system consisting of the set $\cup$, together with the three operations $\wedge, \vee$ and $'$.

We are building an abstract theory in which the elements of $\cup$ can be any objects whatever. In application to propositional logic, $\cup$ will be the set of formulas of propositional logic using only the logical connectives, conjunction, disjunction and negation, and $\wedge, \vee$ and $'$ will be these operations respectively.

In application to Boolean gardens, $\cup$ will be the set of flowers and $\wedge, \vee$ and $'$ will be the operations $\cap, \cup$ and $-$ respectively. [Thus, for any flowers A and B, the flowers $A \wedge B, A \vee B, A'$ will be $A \cap B, A \cup B, \overline{A}$ respectively].

In application to the Boolean theory of sets, $\cup$ will be a collection of subsets of some non-empty set I and $\wedge, \vee$ and $'$ will respectively be intersection, union and complementation.

Now, consider a formula X of propositional logic using no connectives other than $\wedge, \vee$ and $'$. If we now reinterpret the propositional variables p, q, r as names of elements of $\cup$, instead of propositions, and $\wedge, \vee$ and $'$ as the operations meet, join

---

[1]This has been credited to the mathematician E. V. Moore.

and involute of the system B, then X designates some element of $\cup$. For example, if we interpret p as element x of $\cup$, and q as element y, then the formula $p \vee (q \wedge p')$ designates the element $x \vee (y \wedge x')$. [We really went through this twice before for the special cases of Boolean gardens and the Boolean theory of sets]. We now define B to be a *Boolean algebra* or, as we say, $\cup$ is a Boolean algebra with respect to the operations $\wedge, \vee$ and $'$, iff the following condition holds: For any formulas X and Y, using just the connectives $\wedge, \vee$ and $'$, if X is logically equivalent to Y (*i.e.* if $X \equiv Y$ is a tautology), then under *every* interpretation of the propositional variables of X and Y as elements of $\cup$, the formulas X and Y will designate the same element.

Thus, for example, since the formula $(p \wedge q)'$ is logically equivalent to $p' \vee q'$, then in a Boolean algebra, for any elements x and y of $\cup$, it follows that $(x \wedge y)' = x' \vee y'$.

We have already dealt with two examples: In Chapter 17 we saw that in a *reduced* Boolean garden, logically equivalent formulas will always designate the same flower under a given interpretation. Thus, we have

Proposition 1 – A reduced Boolean garden is a Boolean algebra with respect to the operations $\cap, \cup$ and $-$.

We also saw in the last chapter that if $\cup$ is the set of all subsets of some non-empty set I, then logically equivalent formulas always designate the same set under a given interpretation. Thus, we have

Proposition 2 – For any non-empty set I, the collection of all subsets of I is a Boolean algebra with respect to intersection, union and complementation.

Thus, anything we prove about Boolean algebras in general will apply both to reduced Boolean gardens and to the Boolean theory of sets.

It has sometimes been said that the set of formulas of propositional logic, using the connectives $\wedge, \vee$ and $'$ is a Boolean algebra (with respect to conjunction, disjunction and negation). This is not quite true; logically equivalent formulas don't have to be identical! However, there is an important modification of this which *is* true and involves a generalization of Boolean algebras to which we now turn.

Let us go back to considering a general set $\cup$ with operations $\wedge, \vee$ and $'$. Now, let us also consider some relation R between elements x and y of $\cup$. For Boolean gardens, the relation R that will interest us is *similarity* (same color on all days). For propositional logic, we will take R to be the relation of *logical equivalence*. For Boolean set theory, we will take R to be the relation of identity. We shall now define $\cup$ (together with the operation $\wedge, \vee$ and $'$) to be a Boolean algebra *relative to R* iff the following condition holds: For any formulas X and Y, using just the connectives $\wedge, \vee$ and $'$, if X is logically equivalent to Y, then under every interpretation of the propositional variable of X and Y as elements of $\cup$, the element designated by X *bears the relation R* to the element designated by Y. Thus, what we previously called a *Boolean algebra* is under this generalized definition, a Boolean algebra *relative to identity*.

Then it certainly is true that the set of formulas of propositional logic using $\wedge, \vee$ and $'$ is a Boolean algebra *relative to logical equivalence*! It is also true that a Boolean garden, whether reduced or not, is a Boolean algebra relative to the *relation of similarity*. This justifies us in calling Boolean gardens "Boolean".

Boolean Equations – Just as in propositional logic, we use letters p, q, r, with or without subscripts, to stand for arbitrary propositions, in Boolean algebra, we shall use letter x, y, z, w, with or without subscripts, called *variables* to stand for arbitrary elements of $\cup$. By a (Boolean) *term*, we shall mean any expression formed by starting with variables and combining any terms already constructed with the symbols $\wedge, \vee$ and $'$ and using parentheses to avoid ambiguity. Thus,

(1) Each variable is a term.

(2) For any term t, the expressions t' is a term.

(3) For any terms t_1 and t_2, the expressions $(t_1 \wedge t_2)$ and $(t_1 \vee t_2)$ are terms.

It is understood that no expression is a term unless its being so follows from (1), (2) and (3) above. So, for example, $((x \wedge y)' \vee (z \vee (x \wedge y')))'$ is a term.

By a (Boolean) *equation* we mean an expression of the form $t_1 = t_2$, where t_1 and t_2 are terms. For example, $(x \wedge y) = (x \vee y)'$ is an equation (not valid, as it happens). We say that an equation *holds* in a Boolean algebra if it is true for all possible values of the variables. For example, to say that the equation $(x \wedge y)' = (x' \vee y')$ holds in a given Boolean algebra $\cup$, is to say that for *all* elements x and y of $\cup$, the element $(x \wedge y)'$ is the same as the element $x' \vee y'$. We shall call an equation *valid* or an *identity* if it holds for *all* Boolean algebras. Now, we shall call an equation *tautological* if by replacing the Boolean variables by propositional variables and "=" by "≡", we get a tautology of propositional logic. For example, the equation $(x \wedge y)' = x' \vee y'$ is tautological, because $(p \wedge q)' \equiv (p' \vee q')$ is a tautology. Thus, to say that $\cup$ is a Boolean algebra is to say that all tautological equations hold in $\cup$. And so, all tautological equations are valid (they hold in all Boolean algebras).

The Elements 1 and 0 – The equation $x \vee x' = y \vee y'$ is tautological (since $(p \vee p') \equiv (q \vee q')$ is a tautology), and so, in a Boolean algebra, for any element x and y of $\cup$, the element $x \vee x'$ is the same as $y \vee y'$, and we call this common element "1".

Any two logically contradictory formulas are logically equivalent, and so $(p \wedge p') \equiv (q \wedge q')$ is a tautology, hence in a Boolean algebra, for any elements x and y of $\cup$, the element $x \wedge x'$ is the same as the element $y \wedge y'$, and we call this common element "0".

We now add "1" and "0" to our list of terms, and they correspond respectively to the constants T and F of propositional logic. And so, a Boolean equation involving 1 and 0 is to be regarded as tautological if it becomes a tautology, replacing 1 by T and 0 by F, and also making the same replacements as before (= by ≡

and Boolean variables by propositional variables). Thus, for example, $x \wedge 1 = x$ is tautological, since $(p \wedge T) \equiv p$ is a tautology, and so in a Boolean algebra, $x \wedge 1$ is the same element as x, for any x. Also, $T' \equiv F$ and $F' \equiv T$ are tautologies, and so in a Boolean algebra, $1' = 0$ and $0' = 1$. [Let us note that in the Boolean algebra of all subsets of some non-empty set I, the element 0 is the empty set $\emptyset$ and 1 is the set I. Of course, $\overline{I} = \emptyset$ and $\overline{\emptyset} = I$].

Other Boolean Operations – In what follows, B is assumed to be a Boolean algebra.

For any elements x and y of $\cup$, we define $x \supset y$ to be $x' \vee y$. [We recall that in propositional logic, $p \supset q$ is equivalent to $p' \vee q$]. And we define $x \equiv y$ to be the element $(x \supset y) \wedge (y \supset x)$ which is $(x' \vee y) \wedge (y' \vee x)$. [We recall that in propositional logic $p \equiv q$ is logically equivalent to $(p \supset q) \wedge (q \supset p)$. Also, $p \equiv q$ is logically equivalent to $(p \wedge q) \vee (p' \wedge q')$, so in a Boolean algebra, $(x \supset y) \wedge (y \supset x) = (x \wedge y) \vee (x' \wedge y')$].

Thus, for any formulas X and Y, if, under a given interpretation, X and Y designate elements x and y respectively, then $X \supset Y$ will designate $x \supset y$ and $X \equiv Y$ will designate $x \equiv y$.

In accordance with propositional logic, we could define $x \neq y$ as $(x \equiv y)'$, but we rather use the symbol "+" instead of "$\neq$" for Boolean algebras, because it happens to be standard in the literature (and also behaves in certain ways like addition in arithmetic). And so we define $x + y$ to be $(x \equiv y)'$. We could also define $x - y$ to be $x \wedge y'$ (as we did for sets), in which case we could alternatively define $x + y$ to be $(x - y) \vee (y - x)$, which in a Boolean algebra is the same element as $((x \supset y) \wedge (y \supset x))'$, as the reader can verify by a truth table. In the literature, $x + y$ is sometimes called the *symmetric difference* of x and y (since $x + y = (x - y) \vee (y - x)$).

We are now free to add the new symbols $\supset, \equiv$ and + for Boolean algebras, and now any equation, even possibly involving these new symbols, will be called *tautological* if it becomes a tautology by replacing = by $\equiv$, 1 by T, 0 by F, + by $\neq$ and Boolean variables x, y, z, ... by propositional variables p, q, r ....

Let us remark that since $(p \neq p) \equiv F$ is a tautology, then $x + x = 0$ holds in Boolean algebra. In particular, taking 1 for x, we get the strange-looking equation $1 + 1 = 0$, which has no counterpart in arithmetic and is quite baffling to those who have heard of it, but know nothing about Boolean algebra.

Problem 1 – Prove that in a Boolean algebra, $1 \neq 0$ (assuming that $\cup$ has at least 2 elements).

Problem 2 – Prove that in a Boolean algebra, if $x \wedge y = 0$ and $x \wedge y' = 0$, then $x = 0$.

In what follows, B is assumed to be a Boolean algebra.

Problem 3 – Prove that if $x \wedge y = 1$, then $x = 1$ and $y = 1$.

We now define $x \leq y$ if and only if $x \wedge y = x$.

Problem 4 – Prove that $x \leq y$ if and only if $x \supset y = 1$.

Problem 5 – Prove that if $x \leq y$ and $y \leq x$, then $x = y$.

Problem 6 – Prove that $(x \equiv y) = 1$ if and only if $x = y$.

II – Exact Boolean Algebras

We will call a set E of elements of ∪ an *exact* set if for any elements x and y of ∪, the following three conditions hold:

E_1: One of the elements x, x' is in E and the other is not.

E_2: $x \wedge y$ is in E if and only if x and y are both in E.

E_3: $x \vee y$ is in E if and only if at least one of the elements x, y is in E.

Examples – (1) In a Boolean garden, the set of all flowers that are blue on a given day, is an exact set.

(2) In propositional logic, the set of all formulas that are true under a given interpretation is an exact set.

(3) Consider a set I containing at least one element and the set ∪ of all subsets of I. For any element x of I, the set of all subsets of I that contain x is an exact set.

Problem 7 – Suppose ∪ is a Boolean algebra (with respect to $\wedge$, $\vee$ and $'$) and that E is a subset of ∪ which obeys conditions E_1 and E_2 of the definition of an exact set. Show that it also obeys condition E_3. Show that it also obeys the following two conditions:

E_4: $x \supset y$ is in E if and only if either x is outside E or y is inside E.

E_5: $x \equiv y$ is in E if and only if x and y are either both inside E or both outside E.

Problem 8 – Show that in a Boolean algebra, 1 belongs to all exact sets and 0 belongs to none.

We now define a Boolean algebra to be an *exact* Boolean algebra if for all elements x and y, if x and y belong to precisely the same exact sets, then $x = y$.

Examples (1) Consider a reduced Boolean garden. Suppose flowers F and G belong to the same exact sets. Now, given any day, the set of flowers that are blue on that day is an exact set. Therefore, F and G are blue on exactly the same days, hence F = G (since the garden is reduced). Thus, a reduced Boolean garden is an *exact* Boolean algebra.

(2) Suppose I is a set having at least two elements and that $\cup$ is the set of all subsets of I. Suppose that A and B are subsets of I that belong to exactly the same exact sets. For any element x of I, let $\boxtimes$ be the set of all subsets of I that contain x as an element. Then $\boxtimes$ is an exact set, and therefore A belongs to $\boxtimes$ if and only if B belongs to $\boxtimes$, which means that A contains x if and only if B contains x. This is true for *every* element x of I, and so A and B contain exactly the same elements, hence are identical. Thus, $\cup$ is an *exact* Boolean algebra.

As a matter of fact, it can be proved that *every* Boolean algebra is an exact Boolean algebra, but the proof is beyond the scope of this book. We will, however, prove that every *finite* Boolean algebra (every Boolean algebra in which the set $\cup$ has only finitely many elements) is an exact Boolean algebra (a fact that will be quite useful in the next chapter). The proof of this involves several preliminary problems, to which we now turn.

Problem 9 – Suppose that $\cup$ is a Boolean algebra and that 1 is the only element that belongs to all exact sets. Prove that $\cup$ is then an exact Boolean algebra.

Passable Sets – In what follows, $\cup$ is assumed to be a Boolean algebra (with respect to $\wedge, \vee$ and $'$).

By $x_1 \wedge x_2 \wedge x_3$ we shall mean $(x_1 \wedge x_2) \wedge x_3$. By $x_1 \wedge x_2 \wedge x_3 \wedge x_4$ we shall mean $(x_1 \wedge x_2 \wedge x_3) \wedge x_4$, which is $((x_1 \wedge x_2) \wedge x_3) \wedge x_4$, and so forth for any $n \succeq 2$. Thus, $x_1 \wedge \ldots \wedge x_{n+1} = (x_1 \wedge \ldots x_n) \wedge x_{n+1}$. By virtue of the Boolean tautology $x \wedge y = y \wedge x$ and $(x \wedge y) \wedge z = x \wedge (y \wedge z)$, the order in which we arrange the elements $x_1 \wedge \ldots x_n$ makes no difference to the element $x_1 \wedge \ldots \wedge x_n$–that is, if $y_1, \ldots y_n$ are the same elements as $x_1, \ldots x_n$, but perhaps in a different order, then $x_1 \wedge \ldots \wedge x_n = y_1 \wedge \ldots \wedge y_n$. [For example $x \wedge y \wedge z = y \wedge z \wedge x$, because $x \wedge y \wedge z = (x \wedge y) \wedge z = (y \wedge x) \wedge z = y \wedge (x \wedge z) = y \wedge (z \wedge x) = (y \wedge z) \wedge x = y \wedge z \wedge x$].

Given a non-empty set S of elements $x_1, \ldots, x_n$, by the *meet* of S we shall mean the element $x_1 \wedge \ldots \wedge x_n$. [The order in which we take the elements $x_1, \ldots, x_n$ makes no difference, as we have seen]. Thus, for example, if $S = \{x, y, z, w\}$ (the set whose elements are x, y, z and w) then the meet of S is the element $((x \wedge y) \wedge z) \wedge w$. [We remark that in a Boolean algebra of sets (with $\cap, \cup$ and $'$ as the Boolean operations) the meet of a non-empty collection C of sets is simply the *intersection* of all the members of C–*i.e.* the set of all elements that belong to all the sets in C]. We shall henceforth write $\hat{S}$, or $\wedge(S)$ to mean the meet of S.

And now we define a non-empty set S of elements of $\cup$ to be *passable* if $\hat{S} \neq 0$. [We remark that in a reduced Boolean garden, a passable set is simply a non-empty set S of flowers such that there is at least one day on which all the flowers in S are blue].

Problem 10 – Prove that if S is passable, then every non-empty subset of S is passable. (In other words, if some non-empty subset of S is not passable, then S, is not passable).

Problem 11 – Suppose that S is passable. Show that for any elements x and y, the following must be true:

(a) S cannot contain both x and x'.

(b) If S contains $x \wedge y$, then it cannot contain either x' or y'.

(c) If S contains x and y, then it cannot contain $(x \wedge y)'$.

For any set S of elements of ∪, and for any element x of ∪, by S_x, we shall mean the set whose elements are those of S together with the element x. [If x happens to be in S, then S_x, is simply S, but if x is not in S, then S_x has just one more element than S–namely, the element x]. Let us note that regardless of whether x is in S or not, $\hat{S}_x = \hat{S} \wedge x$. [If x is in S, then $\hat{S}_x = \hat{S} \wedge x$ by virtue of the Boolean tautology $x \wedge x = x$].

Problem 12 – Suppose S is passable. Prove that for any element x of ∪, either S_x or $S_{x'}$ is passable.

We call a set S a *maximal passable* set if S is passable and S is not a subset of any other passable set. Now, in a *finite* Boolean algebra, any passable set S must be a subset of some *maximal* passable set, because if S is already a maximal passable set, we are done. If not, then S is a subset of some larger passable set S_1. If S_1 is maximal, we are done, but if not, then S_1 is a subset of some still larger passable set S_2, and so forth. But since ∪ is only finite, we can't go on forever taking larger and larger sets; we must eventually reach a passable set S_n that can't be extended further to a passable set, hence S_n must be a maximal passable set. We record this as

Proposition 3 – In a finite Boolean algebra, every passable set is a subset of some *maximal* passable set.

Let us now note that if M is a maximal passable set, then for any element x, if M_x is passable, then x must be in M, because if x is not in M, M_x is a passable set larger than M, contrary to the fact that M is a *maximal* passable set. Let us record this as

Proposition 4 – If M is a maximal passable set, then for any element x, if M_x is passable, then x is a member of M.

Now we come to the "key" proposition.

Proposition 5 – In a Boolean algebra, any maximal passable set is an exact set.

Problem 13 – Prove Proposition 5.

From Proposition 3 and Proposition 5, we at once have

Theorem 1 – In a finite Boolean algebra, every passable set is a subset of an exact set.

Remark – Theorem 1 also holds if we delete "finite", but the proof of this goes beyond the scope of this book.

Problem 14 – (a) Prove that if $x \neq 0$, then x belongs to at least one exact set (assuming that $\cup$ is finite).
(b) Now prove Theorem 2 below. [Hint: Take advantage of Problem 9].

Theorem 2 – [Main Result] – Every finite Boolean algebra is an exact Boolean algebra.
As we have remarked, the above result will be quite useful in the next chapter.

SOLUTIONS

1 – The following equations are Boolean tautologies.

(1) $x \wedge 1 = x$

(2) $x \wedge 0 = 0$

Now, suppose $1 = 0$. Then for any element x of $\cup$, $x \wedge 1 = x \wedge 0$. Hence by (1) and (2), x = 0. Thus, x = 0 for every element x, which means that 0 is the only element of $\cup$, contrary to our assumption that $\cup$ has at least two elements.

2 – Here are the Boolean tautologies we now need:

(1) $0 \vee 0 = 0$

(2) $y \vee y' = 1$

(3) $x \wedge 1 = x$

(4) $(x \vee y) \wedge (x \vee z) = x \wedge (y \vee z)$

Now, suppose that $x \wedge y = 0$ and $x \wedge y' = 0$. Then $(x \wedge y) \vee (x \wedge y') = 0 \vee 0 = 0$. Also $(x \wedge y) \vee (x \wedge y') = x \wedge (y \vee y')$ (by (4), taking y' for z) $= x \wedge 1$ (by (2), since $y \vee y' = 1$) $= x$ (by (3)). Thus, $(x \wedge y) \vee (x \wedge y')$ equals both 0 and x, hence x = 0.

3 – The Boolean tautologies we need for the solution are the following:

(1) $x \wedge (y \wedge z) = (x \wedge y) \wedge z$

(2) $x' \wedge x = 0$

(3) $0 \wedge y = 0$

(4) $x \wedge 1 = x$

(5) $0' = 1$

(6) $x'' = x$

(7) $x \wedge y = y \wedge x$

Now suppose $x \wedge y = 1$. Then, $x' \wedge (x \wedge y) = x' \wedge 1$.

Hence, $x' \wedge (x \wedge y) = x'$	(since $x' \wedge 1 = x'$ by (4)).
Hence, $(x' \wedge x) \wedge y = x'$	(by (1))
Hence, $0 \wedge y = x'$	(by (2))
Hence, $0 = x'$	(by (3))
Hence, $0' = x''$	
Hence, $1 = x$	(by (5) and (6))

Thus, $x = 1$. Now, $x \wedge y = y \wedge x$ (by (7)), and since $x \wedge y = 1$, then $y \wedge x = 1$. Then, by the same argument (interchanging x with y), $y = 1$.

4 – The following three equations are Boolean tautologies that can be verified by truth tables:

(1) $(x \wedge y) \supset y = 1$

(2) $x \wedge (x \supset y) = x \wedge y$

(3) $x \wedge 1 = x$

Now, we are to show that $x \le y$ if and only if $x \supset y = 1$.

(a) Suppose $x \le y$. Thus, $x \wedge y = x$. Then $x \supset y = (x \wedge y) \supset y$, but $(x \wedge y) \supset y = 1$ (by (1)), and so $x \supset y = 1$.

(b) Conversely, suppose $x \supset y = 1$. Then, $x \wedge (x \supset y) = x \wedge 1$. Hence, $x \wedge (x \supset y) = x$ (since $x \wedge 1 = x$ by (3)). Therefore, $x \wedge y = x$ (since $x \wedge (x \supset y) = x \wedge y$, by (2)), which means that $x \le y$.

5 – Suppose $x \le y$ and $y \le x$. Thus, $x \wedge y = x$ and $y \wedge x = y$, but $y \wedge x = x \wedge y$ (this is a Boolean tautology) and since $y \wedge x = y$, then $x \wedge y = y$. Thus, $x \wedge y$ is equal to both x and y, hence $x = y$.

6 – (a) $(x \equiv x) = 1$ is a Boolean tautology, hence if $x = y$, then $(x \equiv y) = (x \equiv x)$, and hence $x \equiv y$ is then equal to 1.

(b) Conversely, suppose $(x \equiv y) = 1$. Thus $((x \supset y) \wedge (y \supset x)) = 1$. Then by Problem 3, $(x \supset y) = 1$ and $(y \supset x) = 1$. Then by Problem 4, $x \le y$ and $y \le x$. Then by Problem 5, $x = y$.

7 – For this problem we use the Boolean tautology $x \vee y = (x' \wedge y')'$. Now, suppose set E satisfies conditions E_1 and E_2.

(a) Suppose $x \vee y$ is in E. Then $(x' \wedge y')'$ is in E. Hence, $x' \wedge y'$ is outside E (by E_1). Hence, either x' or y' is outside E (because if they were both inside E, then by $E_2, x' \wedge y'$ would be in E, which it isn't). Hence, either x or y must be in E (by E_1). Thus, if $x \vee y$ is in E, then at least,one of the elements x or y (or both) must be in E.

(b) Conversely, suppose that at least one of the elements x, y is in E. Then by E_1 at least one of the elements x', y' must be outside E, hence $x' \wedge y'$ is outside E (because if it were inside E, then x' and y' would both be in E (by E_2), contrary to the fact that at least one of x', y' is outside E). Since $x' \wedge y'$ is outside E, then by $E_1, (x' \wedge y')$, is in E. Thus, $x \vee y$ is in E.

By (a) and (b), $x \vee y$ is in E if and only if at least one of the elements x or y is in E. This proves E_3.

The proofs of E_4 and E_5 are left as exercises.

8 – Suppose E is an exact set. Consider any element x of $\cup$. By condition E_1, at least (in fact exactly) one of the elements x, x' belongs to E and, in either case, $x \vee x'$ belongs to E (by condition E_3), but $x \vee x' = 1$, hence 1 belongs to E. Thus $0'$ belongs to E, hence by $E_1, 0$ doesn't belong to E. And so 1 belongs to E, but 0 doesn't.

9 – Suppose that 1 is the only element that belongs to all exact sets. Now, it follows from E_5 (See Problem 7) that x and y belong to the same exact sets if and only if $x \equiv y$ belongs to *all* exact sets.

Suppose that x and y belong to the same exact sets. Then $x \equiv y$ belongs to *all* exact sets. But since 1 is the only element that belongs to all exact sets, then $(x \equiv y) = 1$. Hence, $x = y$ (Problem 6). Thus $\cup$ is an exact Boolean algebra.

10 – Suppose A is a non-empty subset of S and A is not passable. If A = S, then of course S is not passable, so suppose $A \neq S$. Let $a_1, \ldots, a_n$ be the elements of A and $b_1, \ldots, b_n$ be the elements of S that are not in A. Let $a = a_1 \wedge \ldots \wedge a_n$ and $b = b_1 \wedge \ldots \wedge b_n$. Then $\hat{A} = a$ and $\hat{S} = a \wedge b$. Since A is not passable, then $a = 0$. Hence $a \wedge b = 0$ (since $a \wedge b = 0 \wedge b$ and $0 \wedge b = 0$), and thus $\hat{S} = 0$, so S is not passable.

11 – The following are Boolean tautologies:

(1) $x \wedge x' = 0$

(2) $(x \wedge y) \wedge x' = 0$

(3) $(x \wedge y) \wedge y' = 0$

(4) $x \wedge y \wedge (x \wedge y)' = 0$

Therefore, none of the following sets are passable:

(1) $\{x, x'\}$

(2) $\{x \land y, x'\}$

(3) $\{x \land y, y'\}$

(4) $\{x, y, (x \land y)'\}$

Then, by Problem 10, none of these four sets can be subsets of a passable set S.

12 – Suppose that neither S_x nor $S_{x'}$ is passable. Thus $\hat{S}_x = 0$ and $\hat{S}_{x'} = 0$. Let $s = \hat{S}$. Then $\hat{S}_x = s \land x$ and $S'_x = s \land x'$. Thus $s \land x = 0$ and $s \land x' = 0$. Hence, $s = 0$ (Problem 2) and so, S is not passable.

Thus, if neither S_x nor $S_{x'}$ is passable, then S is not passable. Therefore, if S *is* passable, then at least one of the sets S_x or $S_{x'}$ must be passable.

13 – Suppose M is a maximal passable set.

(1) By Problem 12, either M_x or $M_{x'}$ is passable. Then by Proposition 4, either x or x' is in M. Also, x and x' cannot both be in M (Problem 11 (a)). Thus, one and only one of the elements x, x' is in M, and so condition E_1 (of the definition of an exact set) is fulfilled.

(2) As for condition E_2, suppose $x \land y$ is in M. Then M contains neither x' nor y' (Problem 11 (b)), hence M contains both x and y (by condition E_1, which we have seen to hold for M).

Also, suppose M contains both x and y. Then M cannot contain $(x \land y)'$ (Problem 11 (c)), hence by E_1, M contains $x \land y$. Thus, M contains $x \land y$ if and only if M contains both x and y, hence M satisfies condition E_2. Since E_1 and E_2 are satisfied, so is E_3 (Problem 7), and so M is an exact set.

14 – (a) Suppose $x \neq 0$. Let S be the set $\{x\}$ (the set whose only member is x). Then $\hat{S} = x$, so $\hat{S} \neq 0$, hence S is passable. Then by Theorem 1, S is a subset of some exact set E, and so x is a member of E.

(b) By Problem 9, it suffices to show that 1 is the only element that belongs to all exact sets. Well, suppose x is any element other than 1. Then $x' \neq 0$ (because if x' were 0, we would have $x'' = 0'$, and hence x would be 1), hence by (a), x' belongs to some exact set E. Then x is not in E (by condition E_1) and so x does not belong to all exact sets. Thus 1 is the only element that belongs to all exact sets.

Chapter 20

Boolean Gardens Revisited

All the results in Chapters 15 and 16 about reduced Boolean gardens and Boolean islands of variable liars are now easily generalizable to finite Boolean algebras in general–particularly easy now that we know that every finite Boolean algebra is an exact Boolean algebra. Roughly speaking, "exact" sets correspond to "days". We recall that we said that a flower G *dominates* a flower F if G is blue on all days that F is blue. Well, more generally, given a finite Boolean algebra, we will say that an element y *dominates* an element x if y belongs to all exact sets which contain x. In the solutions to the problems of Chapter 15, we proved (Proposition 1) that flower F dominates flower G if and only if $G \wedge F = G$. Well, only a slight verbal change in the proof is necessary to establish that in an exact Boolean algebra, an element y dominates element x if and only if $x \wedge y = x$. Thus, y dominates x if and only if $x \leq y$.

Remarks – (1) For the Boolean algebra of sets, $A \leq B$ simply says that $A \subseteq B$, since $A \wedge B = A$ is equivalent to $A \subseteq B$. Thus, B dominates A if and only if A is a subset of B.

(2) We might note an analogous thing in propositional logic: $(p \wedge q) \equiv p$ is equivalent to $p \supset q$. Thus, q dominates p if and only if p implies q.

Atomic Elements – For Boolean algebra in general, we might call x a *basic* element if x dominates no element other than x and 0, and if $x \neq 0$. However, the established term for such an x is an *atomic element*, or a *Boolean atom*, or more briefly, an *atom*; and so, we had best use the established terminology.

As with flower gardens, we will say that x *represents* the set of all atoms that x dominates.

We have already defined the *meet* $x, \wedge \ldots \wedge x_n$ of a finite set $\{x, \ldots, x_n\}$, and we note that this element $x_1 \wedge \ldots \wedge x_n$ is the one and only element that belongs to all exact sets that contain all of the elements $x_1, \ldots, x_n$. We also define the *join* of $\{x_1, \ldots, x_n\}$ as the one and only element that belongs to all exact sets that contain at least one of the elements $x_1, \ldots, x_n$ (as we did for Boolean gardens in Chapter 15). We note that the join of $\{x_1, x_2\}$ is $x_1 \vee x_2$; the join of $\{x_1, x_2, x_3\}$ is $(x_1 \vee x_2) \vee x_3$, which we more simply write $x_1 \vee x_2 \vee x_3$. More generally, the join of $\{x_1, \ldots, x_n\}$ is written $x_1 \vee \ldots \vee x_n$. We note that for any $n \geq 2$, the element $x_1 \vee \ldots \vee x_{n+1} = (x_1 \vee \ldots \vee x_n) \vee x_{n+1}$.

As with Boolean gardens discussed in Chapter 15, for any element x, we let x^* be the set of all *atomic* elements dominated by x–in other words, the set *represented* by x. The propositions 1-10 proved in the solutions of Chapter 15 are easily generalizable to the following respective facts about *finite* Boolean algebras.

F_1: y dominates x iff $x \wedge y = x$ *(i.e.* iff $x \leq y$), which in turn is the case iff $y' \wedge x = 0$.

F_2: The only element that can be dominated by both x and x' is 0.

F_3: No atomic element can dominate any other atomic element.

F_4: If y fails to dominate x and x is atomic, then $x \wedge y = 0$.

F_5: For any *atomic* element x and any element y, either y dominates x or y' dominates x.

F_6: Two distinct atomic elements x and y cannot be members of the same exact set.

F_7: For any atomic element x, there is an exact set E that contains x and no other atomic element.

F_8: Every element other than 0 dominates some atomic element.

F_9: If x and y are distinct elements, then one of them dominates some atomic element which is not dominated by the other.

F_{10}: Every set S of atomic elements is represented by the join of S.

Propositions A, B, C, and Theorem 1 of Chapter 15 have the following respective generalizations to finite Boolean algebra.

F_{11}: If $x \neq y$ then $x^* \neq y^*$.

F_{12}: A set of atomic elements cannot be represented by more than one element.

F_{13}: A set S of atomic elements is represented by one and only one element– namely, the join of S.

F_{14}: There is a 1-1 correspondence between the elements of $\cup$ and the sets of atomic elements–namely, the correspondence in which each element x of $\cup$ corresponds to the set x^* of all atomic elements dominated by x.

From F_{14} we have the following fundamental and famous result about finite Boolean algebras.

Theorem 1 – In a finite Boolean algebra, the number of elements is the same as the number of sets of atomic elements–this number being 2^n, where n is the number of atomic elements.

Now, let us consider generalizations of some of the things proved in Chapter 16. We proved (F_3 of Chapter 16) that on each day, exactly one basic flower is blue and all the other basic flowers are red, and that for each basic flower X, there is a day an which X is blue and all the other basic flowers are red. This generalizes to the following fact about finite Boolean algebras.

F_{15}: Each exact set contains one and only one atomic element, and each atomic element belongs to some exact set.

Fact F_4 of Chapter 16 (which we recall is that the entire color configuration on a given day is completely determined by which basic flower is blue on that day) is but a special case of the following fact about finite Boolean algebras.

F_{16}: An atomic element x is a member of only one exact set.

From F_{15} and F_{16} we have

F_{17}: In a finite Boolean algebra, the number of exact sets is the same as the number of atomic elements.

In Chapter 16, we showed (F_6) that if $A \cup B$ dominates X and X is basic, then either A dominates X or B dominates X. This generalizes to the following fact about finite Boolean algebras.

F_{18}: If $x \vee y$ dominates z and z is *atomic*, then either x dominates z or y dominates z (or both).

Of particular importance is the following generalization of Fact F_7 of Chapter 16.

F_{19}: For any elements x and y of a finite Boolean algebra:

(1) $(x')^* = (x^*)'$

(2) $(x \vee y)^* = x^* \vee y^*$

(3) $(x \wedge y)^* = x^* \wedge y^*$

Boolean Isomorphisms – Suppose now that $\cup_1$ is a Boolean algebra with respect to operations $\wedge, \vee$ and $'$, and that $\cup_2$ is another Boolean algebra with respect to operations which we will denote by different symbols–say $\cap, \cup$ and $-$, and that $\cup_2$ has the same number of elements as $\cup_1$. Now, consider a 1-1 correspondence between the elements of $\cup_1$ and the elements of $\cup_2$. For each element x of $\cup_1$ let $x^\sharp$ be the corresponding element of $\cup_2$. The correspondence is called a *Boolean isomorphism*, or just an *isomorphism* for short, if for any elements x and y of $\cup_1$, the following three conditions hold:

(1) $(x')^\sharp = \overline{x^\sharp}$

(2) $(x \vee y)^\sharp = x^\sharp \cup y^\sharp$

(3) $(x \wedge y)^\sharp = x^\sharp \cap y^\sharp$

Stated otherwise, the correspondent is a Boolean isomorphism iff for any elements x_1 and y_1 of $\cup_1$ and their respective corresponding elements x_2 and y_2 of $\cup_2$:

(1) x_1' corresponds to $\overline{x_1}$

(2) $x_1 \vee y_1$ corresponds to $x_2 \cup y_2$

(3) $x_1 \wedge y_1$ corresponds to $x_2 \cap y_2$

If there exists a Boolean isomorphism from one Boolean algebra to another, then the two Boolean algebras are called *isomorphic.*

What we called a *Boolean match* in Chapter 16 is thus a Boolean isomorphism from the Boolean algebra of the men on the male Boolean island to the Boolean algebra of the women on the female Boolean island. We proved, in Chapter 16 that if the number of men on the male Boolean island is the same as the number of women on the female Boolean island, then a Boolean match is possible. The purely mathematical content of this is the following important and well-known result:

Theorem 2 – Any two finite Boolean algebras with the same number of elements are isomorphic.

We also showed in Chapter 16 that if n is the number of basic inhabitants of each of the islands, then the number of possible Boolean matches is n! (factorial n). The purely mathematical content of this is the following:

Theorem 3 – Suppose that B_1 and B_2 are finite Boolean algebras with the same number of elements, and that n is the number of atomic elements of B_1 (or of B_2, which is the same). Then, the number of Boolean isomorphisms from B_1 to B_2 is n!

Remarks – All these generalizations of results of Chapters 15 and 16 can be proved in the same manner as the proofs in those chapters, only replacing "set of flowers blue on a given day", or "set of people truthful on a given day" by "exact set". It is not necessary to repeat here these proofs.

Boolean Algebras of Sets – By a *Boolean algebra of sets* is meant a Boolean algebra in which 1 is some non-empty set I, $\cup$ is a set of subsets of I but not necessarily the set of *all* subsets of I) and $\wedge, \vee,'$ are respectively the operations $\cap, \cup, -$ (intersection, union, complementation) and for any elements A and B of $\cup$, the sets $A \cap B, A \cup B$ and $\overline{A}$ are also elements of $\cup$. If $\cup$ happens to contain *all* subsets of I, then we will call $\cup$ a *total* Boolean algebra of sets.

We have already seen that in a total Boolean algebra of sets, the Boolean atoms are simply the subsets of I which contain just one element.

An interesting question arises: Is every Boolean algebra isomorphic to some total Boolean algebra of sets? It is well-known that the answer is *no*, but the proof of this goes beyond the scope of this book (but see *Discussion* below for a reference). However, by Facts 14 and 19, we have the following well-known result:

Theorem 4 – Every *finite* Boolean algebra is isomorphic to a total Boolean algebra of sets–namely, the set of all subsets of the set of all the atomic elements.

Discussion – Since it is not true that every Boolean algebra is isomorphic to a total Boolean algebra of sets, we might try for a second best: Is every Boolean algebra at least isomorphic to some Boolean algebra of sets (but not necessarily total)? Happily, the answer is *yes*, and this famous result is known as Marshal Stone's *representation theorem*. Again, the proof goes beyond the scope of this book, but the interested reader can find a proof in *Rosenbloom*, cited below, in which it is also proved that not every Boolean algebra is isomorphic to a *total* Boolean algebra of sets. For a proof that every Boolean algebra is an exact Boolean algebra, one reference is Bell and Slomson cited below.

References

1 Paul Rosenbloom, *The Elements of Mathematical Logic*, Dover Publications, 1950.

2 J.L. Bell and A.B. Slomson, *Models and Ultraproducts*, North Holland Publishing Company, Amsterdam. Oxford American Elsevier Publishing Company, Inc. New York, 1969.

Chapter 21

Another Grand Problem

I – Statement of the Problem

The following Boolean equations hold in all Boolean algebras, because they are all tautological (they come from the tautologies $T_1 - T_9$ of Chapter 17).

P_1: $x \wedge y = y \wedge x$

P_2: $x \wedge (y \wedge z) = (x \wedge y) \wedge z$

P_3: $x \wedge (y \vee z) = (x \wedge y) \vee (x \wedge z)$

P_4: $(x \vee y)' = x' \wedge y'$

P_5: $x'' = x$

P_6: $x \wedge x' = 0$

P_7: $x \wedge 0 = 0$

P_8: $x \wedge 1 = x$

P_9: $0' = 1$

Now comes a "grand" problem: Suppose we are given a set $\cup$ and operations $\wedge, \vee$ and $'$, without being told that we have a Boolean algebra, but we are told that conditions $P_1 - P_5$ all hold and that $\cup$ contains elements 0 and 1 such that $P_6 - P_9$ also hold. The problem is to prove that we then have a Boolean algebra.

The amazing thing, then, is that from just the nine tautological equations above, one can derive *all* tautological equations (of which there are infinitely many!). To prove this, is indeed a "grand" undertaking, in that a host of smaller problems must be solved first.

In technical terms, a set of conditions about the operations $\wedge, \vee$ and $'$ is called an *adequate postulate system* (for Boolean algebras) if the conditions all hold for all Boolean algebras, and if any system satisfying the conditions is a Boolean algebra.

Our purpose, then, is to show that laws $P_1 - P_9$ constitute an adequate postulate system for Boolean algebras. Two other postulate systems will emerge along the way, and another postulate system will also be shown to be adequate.

There are indeed many, many adequate postulate systems for Boolean algebras that have appeared in the literature, and the study of their interrelationships (known as *postulate theory*) is a most fascinating subject. This chapter is a bit technical, though, not very difficult. By contrast, the next chapter is essentially "newsy" and can be read independently of this one, if desired.

Comments – Before turning to the solution of the problem, we wish to acquaint the reader with some standard terminology.

Consider an operation that assigns to any element x of a set S and any element y of S some element of S denoted $x * y$. The operation is called *commutative* if $x * y = y * x$ (for all elements x and y of S). P_1 above tells us that the operation $\wedge$ is commutative. [Familiar commutative operations in arithmetic are "plus" and "times": $x + y = y + x$ and $xy = yx$]. An operation $*$ is called *associative* if $x * (y * z) = (x * y) * z$. Law P_2 says that $\wedge$ is associative. [In arithmetic, "plus" and "times" are associative, since $x + (y + z) = (x + y) + z$ and $x(yz) = (xy)z$]. Law P_3 is called a *distributive* law and is paraphrased by saying that $\wedge$ distributes with $\vee$. [In arithmetic, "times" distributes with "plus", since $x(y + z) = xy + xz$, but "plus" does *not* distribute with "times", since in general it is not true that $x + yz = (x + y)(x + z)$]. Law P_4 is one of the two laws call *De Morgan's laws*. Law P_5 is sometimes called the *involution law*.

II – Solution of the Problem

We now turn to the proof that the conditions $P_1 - P_9$ constitute an adequate postulate system for Boolean algebras. We will show how *every* tautological Boolean equation can be derived from just these nine. We must prove many subsidiary results along the way.

In deriving other equations from $P_1 - P_9$, it is to be understood that once two terms t_1 and t_2 have been proved equal, then we can conclude that $t_{1'} = t_{2'}$, and that for any term t, we can conclude $t_1 \wedge t = t_2 \wedge t$, and that $t \wedge t_1 = t \wedge t_2$, and that $t_1 \vee t = t_2 \vee t$, and $t \vee t_1 = t \vee t_2$. In brief, "equals may be substituted for equals".

As a second principle, once an equation has been proved, then we can substitute any terms we like for the variables. For example, P_1 says that for *any* elements x and y, $x \wedge y = y \wedge x$. Hence, we can conclude, say, that $(x \vee z') \wedge (w \wedge y) = (w \wedge y) \wedge (x \vee z')$. We have simple substituted $x \vee z'$ for x and $w \wedge y$ for y in the equation $x \wedge y = y \wedge x$.

And now for some preliminary problems.

Problem 1 – From $P_1 - P_9$, derive $P_1^\circ - P_9^\circ$ below.

P_1°: $x \vee y = y \vee x$

P_2°: $x \vee (y \vee z) = (x \vee y) \vee z$

P_3°: $x \vee (y \wedge z) = (x \vee y) \wedge (x \vee z)$

P_4°: $(x \wedge y)' = x' \vee y'$

P_5°: $x'' = x$ (same as P_5)

P_6°: $x \vee x' = 1$

P_7°: $x \vee 1 = 1$

P_8°: $x \vee 0 = x$

P_9°: $1' = 0$

Duality – By the *dual* of a Boolean equation (involving only the operations $\wedge$, $\vee$ and $'$ and the constants 1 and 0) is meant the result of interchanging $\wedge$ with $\vee$ and 1 with 0. We note that in the above laws $P_1^\circ - P_9^\circ$, the law P_1° is the dual of P_1, P_2° is the dual of P_2, and so forth down to P_9° is the dual of P_9. For any equation E, we denote its dual by E°. Now, if E is any equation derivable from $P_1 - P_9$, then its dual E° is derivable from $P_1^\circ - P_9^\circ$ by the same argument, only interchanging $\wedge$ with $\vee$ and 1 with 0, and since $P_1^\circ - P_9^\circ$ are themselves derivable from $P_1 - P_9$, then E° is derivable from $P_1 - P_9$. We thus have the following useful principle:

Proposition 1 – [A Duality Principle] – The dual of any equation derivable from $P_1 - P_9$ is also derivable from $P_1 - P_9$.

This duality principle cuts our future labor in half, since once we have derived an equation E (from $P_1 - P_9$) we now get its dual E° free (though in a sense, we have already paid for it by deriving $P_1^\circ - P_9^\circ$ in the first place).

We note that we could have alternatively started with the postulates $P_1^\circ - P_9^\circ$ and derived $P_1 - P_9$ from them by dual arguments (same arguments, only interchanging $\wedge$ with $\vee$ and 1 with 0). And so once we have proved that $P_1 - P_9$ is an adequate postulate system, it will follow that $P_1^\circ - P_9^\circ$ is also.

We now turn to another duality principle. We have so far derived only equations from $P_1 - P_9$, but we will shortly be deriving various *relations* between equations such as "if equation E_1 holds, so does E_2", or perhaps "E_1 is equivalent to E_2". Now, suppose that from $P_1 - P_9$ we have proved that E_1 implies E_2. Then by interchanging $\wedge$ with $\vee$ and 1 with 0, we could in the same way prove from $P_1^\circ - P_9^\circ$ that E_1° implies E_2°, and since $P_1^\circ - P_9^\circ$ are all derivable from $P_1 - P_9$ follows that E_1° implies E_2°. And so if E_1 implies E_2 is derivable from $P_1 - P_9$, so is E_1° implies E_2°. From this is further follows that if from $P_1 - P_9$ we can prove E_1 iff E_2, then from $P_1 - P_9$ we can prove E_1° iff E_2°. And so, we have:

Proposition 2 – [Another Duality Principle] – For any equations E_1 and E_2:

(a) If (E_1 implies E_2) is derivable from $P_1 - P_9$, so is (E_1° implies E_2°).

(b) If (E_1 is equivalent to E_2) is derivable from $P_1 - P_9$, so is (E_1° is equivalent to E_2°).

Now, we wish to derive more laws from $P_1 - P_9$.

Problem 2 – From $P_1 - P_9$ (and any laws already derived from them) prove laws $Q_1 - Q_{15}$ below.

Q_1: $x = y$ iff $x' = y'$

Q_2: (a) $(x \wedge y) \vee (x \wedge y') = x$

   (b) $(x \vee y) \wedge (x \vee y') = x$

Q_3: The following four conditions are all equivalent:

   (1) $x \wedge y = x$

   (2) $x \wedge y' = 0$

   (3) $x' \vee y = 1$

   (4) $x \vee y = y$

Q_4: $x \leq y$ if and only if $y' \leq x'$

Q_5: $x \leq x$ (*i.e.*) $x \wedge x = x$

Q_6: (a) $x \wedge y \leq x$

   (b) $x \wedge y \leq y$

   (c) $x \leq x \vee y$

   (d) $y \leq x \vee y$

Q_7: (a) If $x = y$, then $x \leq y$

   (b) $x \leq x''$

   (c) $x' \wedge y' \leq (x \vee y)'$

Q_8: If $x \leq y$ and $y \leq z$ then $x \leq z$

Q_9: (a) If $x \leq y$ and $x \leq z$ then $x \leq y \wedge z$

   (b) If $x \leq z$ and $y \leq z$ then $(x \vee y) \leq z$

Q_{10}: If $x \wedge y \leq z$ and $x \wedge y' \leq z$ then $x \leq z$

Q_{11}: If $x \leq y$ and $y \leq x$ then $x = y$

Q_{12}: (a) $x \leq 1$

(b) If $1 \leq x$ then $x = 1$

(c) $0 \leq x$

(d) If $x \leq 0$ then $x = 0$

Q_{13}: If $x \leq z$ and $x' \leq z$ then $z = 1$

Q_{14}: $x = y$ if and only if $(x \equiv y) = 1$

Q_{15}: If $x \leq y \equiv z$ and $x' \leq y \equiv z$ then $y = z$

We shall now re-label some of the laws that we have derived from $P_1 - P_9$ as $L_1 - L_{12}$ below (which are respectively Q_5, Q_6(a), Q_6(b), Q_6(c), Q_6(d), Q_7(c), Q_7(b), Q_8, Q_9(a), Q_4, Q_{10} and Q_{15}). As we will see, these laws $L_1 - L_{12}$ themselves constitute another adequate postulate system for Boolean algebras.

L_1: $x \leq x$

L_2: $x \wedge y \leq x$

L_3: $x \wedge y \leq y$

L_4: $x \leq x \vee y$

L_5: $y \leq x \vee y$

L_6: $(x' \wedge y') \leq (x \vee y)'$

L_7: $x \leq x''$

L_8: If $x \leq y$ and $y \leq z$, then $x \leq z$

L_9: If $x \leq y$ and $x \leq z$, then $x \leq y \wedge z$

L_{10}: If $x \leq y$, then $y' \leq x'$, and conversely.

L_{11}: If $x \wedge y \leq z$ and $x \wedge y' \leq z$, then $x \leq z$

L_{12}: If $x \leq y \equiv z$ and $x' \leq y \equiv z$, then $y = z$

The postulates $P_1 - P_9$ we will call the *P-group*. The postulates $L_1 - L_{12}$ we will call the *L-group*. We have derived the L-group from the P-group, and we will show that the L-group is itself an adequate postulate system for Boolean algebras, and hence that the P-group also is. Well, from the L-group, we will derive a third group–$B_1 - B_5$ below–which we will call the *B-group* and which we will show is, in turn, an adequate postulate system for Boolean algebras.

Problem 3 – From $L_1 - L_{12}$ prove laws $B_1 - B_5$ below.

B_1: (a) $x \leq x$

(b) If $x \leq y$, then $x \wedge z \leq y$ and $z \wedge x \leq y$

B_2: (a) If $x \leq y$ and $x \leq z$, then $x \leq y \wedge z$

(b) If either $x \leq y'$ or $x \leq z'$, then $x \leq (y \wedge z)'$

B_3: (a) If either $x \leq y$ or $x \leq z$, then $x \leq y \vee z$

(b) If $x \leq y'$ and $x \leq z'$, then $x \leq (y \vee z)'$

B_4: If $x \leq y$, then $x \leq y''$

B_5: (a) If $x \wedge y \leq z$ and $x \wedge y' \leq z$, then $x \leq z$

(b) If $x \leq y \equiv z$ and $x' \leq y \equiv z$, then $y = z$.

From B_2, B_3 and B_4 easily follow the next two laws.

B_6: (a) If either $x \leq y'$ or $x \leq z$, then $x \leq y \supset z$

(b) If $x \leq y$ and $x \leq z'$, then $x \leq (y \supset z)'$

B_7: (a) If $x \leq y$ and $x \leq z$, then $x \leq y \equiv z$

(b) If $x \leq y$ and $x \leq z'$, then $x \leq (y \equiv z)'$

(c) If $x \leq y'$ and $x \leq z$, then $x \leq (y \equiv z)'$

(d) If $x \leq y'$ and $x \leq z'$, then $x \leq y \equiv z$

Proof – B_6: (a) Suppose that either $x \leq y'$ or $x \leq z$. Then $x \leq y' \vee z$ (by B_3 (a)) ... ie. $x \leq y \supset z$.

(b) Suppose $x \leq y$ and $x \leq z'$. Since $x \leq y$ then $x \leq y''$ (B_4). Thus $x \leq y''$ and $x \leq z'$, hence $x \leq (y' \vee z)'$ (by B_3 (b), taking y' for y) ... i.e. $x \leq (x \supset y)'$.

B_7: (a) Suppose $x \leq y$ and $x \leq z$. Since $x \leq z$, then $x \leq y \supset z$ (by B_6 (a)). Since $x \leq y$, then $x \leq (z \supset y)$ (again by B_6 (a)). Thus $x \leq y \supset z$ and $x \leq z \supset y$, and so $x \leq (y \supset z) \wedge (z \supset y)$ (by B_2 (a)) ... i.e. $x \leq y \equiv z$.

(b) Suppose $x \leq y$ and $x \leq z'$. Then $x \leq (y \supset z)'$ (by B_6 (b)), hence $x \leq ((y \supset z) \wedge (z \supset y))'$ (by B_2 (b)).

(c) If $x \leq y'$ and $x \leq z$, then $x \leq (z \supset y)'$ (again by B_6 (b)), hence again $x \leq ((x \supset y) \wedge (y \supset x))'$ (by B_2 (b)).

(d) Suppose $x \leq y'$ and $x \leq z'$. Since $x \leq y'$, then $x \leq y \supset z$ (B_6 (a)) and since $x \leq z'$, then $x \leq (z \supset y)$, hence $x \leq y \supset z$ and $x \leq z \supset y$, so $x \leq ((y \supset z) \wedge (z \supset y))$ (by B_2 (a)).

Adequacy of the B-group – We now wish to show that the B-group is an adequate postulate system for Boolean algebras (and hence that the L-group and the P-group are also adequate). First, some words about the important principle known as *mathematical induction*:

Suppose I tell you that on a certain planet it is raining today and that on this planet, it never rains one day without raining the next day as well. In other words, on any day on which it rains, it rains the next day as well. Isn't it obvious that it follows that it's going to rain *every day* from now on? This is but a special case of the principle of *mathematical induction*, which is that if something is true for the number 1, and if it is never true for any positive whole number n without being true for $n+1$ as well, then it is true for *all* positive whole numbers (since it is then successively true for 2, 3, 4, 5, ... and so forth).

There is a similar principle for *terms* in Boolean algebra. We recall that terms are constructed starting with *variables* x, y, z, w, with or without subscripts, and forming new terms from old ones by taking t' for any term t already constructed and taking $t_1 \wedge t_2$ and $t_1 \vee t_2$ for any terms t_1 and t_2 already constructed. The induction principle for terms is this: If a given property holds for the variables (which are the simplest possible terms) and the property is such that for any term t for which it holds, it also holds for t', and for any terms t_1 and t_2 for which it holds, it also holds for $t_1 \wedge t_2$ and $t_1 \vee t_2$; then under these conditions, the property holds for *all* terms. [For example, to show that these conditions imply that the property holds for the term $x \wedge (y \vee z)'$, we see that since it holds for the variable y and z, it must also hold for $y \vee z$, and hence also for $(y \vee z)'$, and since it holds also for the variable x, it must hold for $x \wedge (y \vee z)'$].

Let us record these inductive principles as

Proposition 3 – [Principles of Mathematical Induction]

(a) If a property holds for the number 1, and if it never holds for any positive whole number n without also holding for $n + 1$, then the property holds for *all* positive whole numbers.

(b) If a property holds for all Boolean variables, and if it holds for term t' whenever it holds for t, and if it holds for the terms $t_1 \wedge t_2$ and $t_1 \vee t_2$ whenever it holds for both t_1 and t_2, then the property holds for *all* terms.

We now need to establish an important consequence of laws $B_1(a)$ and $B_1(b)$.

B_1^*: For any (positive whole number) n and any number i from 1 to n:
$$x_1 \wedge \ldots \wedge x_n \leq x_i$$

Thus, for example, for $n = 3$, B_1^* says that $x_1 \wedge x_2 \wedge x_3 \leq x_1$ and $x_1 \wedge x_2 \wedge x_3 \leq x_2$, and $x_1 \wedge x_2 \wedge x_3 \leq x_3$–in brief, $x_1 \wedge x_2 \wedge x_3$ is dominated by x_1, x_2 and x_3.

We can see this as follows: By $B_1(a)$, $x \leq x$, and so by $B_1(b)$, $x \wedge y \leq x$ and $y \wedge x \leq x$. Thus, for $n = 2$, $x_1 \wedge x_2 \leq x_1$ and $x_1 \wedge x_2 \leq x_2$. Now, consider the case of three variables x_1, x_2 and x_3. Since $x_1 \wedge x_2 \leq x_1$ and $x_1 \wedge x_2 \leq x_2$, then by $B_1(b)$, $x_1 \wedge x_2 \wedge x_3 \leq x_1$ and $x_1 \wedge x_2 \wedge x_3 \leq x_2$. Also, $x_3 \leq x_3$ (by $B_1(a)$), hence, $x_1 \wedge x_2 \wedge x_3 \leq x_3$. And so, for $i = 1, i = 2$, or $i = 3$, we see that $x_1 \wedge x_2 \wedge x_3 \leq x_i$. A similar argument works for $n = 4, n = 5$, and so forth.

A more rigorous proof uses mathematical induction. For $n = 1$, the statement $x_1 \wedge \ldots \wedge x_n \leq x_1$ simply says that $x_1 \leq x_1$, which is true by $B_1(\text{a})$. Now, suppose that n is any number such that $x_1 \wedge \ldots \wedge x_n \leq x_i$ for each number i from 1 to n. We must show that $n + 1$ is another such number–in other words that $x_1 \wedge \ldots \wedge x_{n+1} \leq x_i$ for each i from 1 to $n + 1$. Well, consider first any number i from 1 to n. Since $x_1 \wedge \ldots \wedge x_n \leq x_i$, then by $B_1(\text{b})$, $(x_1 \wedge \ldots \wedge x_n) \wedge x_{n+1} \leq x_i$–thus $x_1 \wedge \ldots \wedge x_{n+1} \leq x_i$ for each i from 1 to n. But also, $x_{n+1} \leq x_{n+1}$ (by $B_1(\text{a})$) and so $(x_1 \wedge \ldots \wedge x_n) \wedge x_{n+1} \leq x_{n+1}$ (by $B_1(\text{b})$). Thus for each i from 1 to $n + 1$, we see that $x_1 \wedge \ldots \wedge x_{n+1} \leq x_i$.

And so the result is true for $n = 1$, and for any n for which the result is true, the result is also true for $n + 1$. And so, by the principle of mathematical induction, the result is true for *every* positive whole number n. Thus B_1^* is established.

Basic Terms – By a basic term in the variables $x_1 \ldots x_n$, we mean a term $x_1^* \wedge \ldots \wedge x_n^*$, where x_1^* is either x_1 or x_1'; x_2^* is either x_2 or x_2' (but not necessarily respectively), $\ldots x_n^*$ is either x_n or x_n'. [Thus for each i from 1 to n, x_i^* is either x_i or x_i'].

For $n = 2$, there are four basic terms:

$x_1 \wedge x_2$

$x_1 \wedge x_2'$

$x_1' \wedge x_2$

$x_1' \wedge x_2'$

For $n = 3$, there are eight basic terms:

$x_1 \wedge x_2 \wedge x_3$

$x_1 \wedge x_2 \wedge x_3'$

$x_1 \wedge x_2' \wedge x_3$

$x_1 \wedge x_2' \wedge x_3'$

$x_1' \wedge x_2 \wedge x_3$

$x_1' \wedge x_2 \wedge x_3'$

$x_1' \wedge x_2' \wedge x_3$

$x_1' \wedge x_2' \wedge x_3'$

In general, for each n, there are 2^n basic terms.

Given now a basic term $x_1^* \wedge \ldots \wedge x_n^*$ and a term t whose variables are all included in the list $x_1, \ldots, x_n$, we shall say that $x_1^* \wedge \ldots \wedge x_n^*$ *governs* the term t if either $x_1^* \wedge \ldots \wedge x_n^* \leq t$ or $x_1^* \wedge \ldots \wedge x_n^* \leq t'$.

From B_1^*, B_2, B_3 and B_4 follows the following crucial fact about basic terms:

B_8: A basic term $x_1^* \wedge \ldots \wedge x_n^*$ governs *every* term t whose variables are all among $x_1, \ldots, x_n$.

Problem 4 – Using B_1^*, B_2, B_3 and B_4, prove B_8. [Hint: First prove it for the case that t is simply one of the variables $x_1, \ldots, x_n$ (which is really obvious!). Then show that if t_1 and t_2 are so governed, so are $t_1 \wedge t_2$ and $t_1 \vee t_2$, and that if t is a term so governed, so is t'. Conclude, using mathematical induction, that *every* term t in the variables $x_1, \ldots, x_n$ is so governed].

We shall say that a term t is *identically* 0 if it has the value 0 for all values of its variables, and *identically* 1 if it takes on the value 1 for all values of its variables. [For example, the term $x \wedge x'$ is identically 0, since for *every* element x of $\cup$, the element $x \wedge x'$ is 0, whereas the term $x \vee x'$ is identically 1].

Problem 5 – (a) Prove that no basic term is identically 0.
  (b) Prove that for any basic term B and any term t, it is not possible for $B \leq t$ and $B \leq t'$ to both be Boolean tautologies.
  (c) Prove that if B is a basic term and if t is a term whose variables are included in the variables of B, and if t is identically 1, then $B \leq t$ is derivable as a consequence of B_{1^*}, B_2, B_3 and B_4.

Problem 6 – Prove the law B_9 below.

B_9: If $t_1 = t_2$ is a tautological Boolean equation, then for every basic term B whose variables include all those of $t_1 = t_2$, the equation $B \leq t_1 \equiv t_2$ (*i.e.* the equation $B \wedge (t_1 \equiv t_2) = B$) is a consequence of $B_1^*, B_2 - B_4$.

Law B_9 is one basic fact about basic terms that we will need. We now turn to a second basic fact which, with B_9, easily shows that the B-group is an adequate postulate system for Boolean algebras.
  Let us first consider basic terms in just two variables x_1 and x_2. Suppose that t_1 and t_2 are terms such that the following four equations are provable from the B-group.

(1) $x_1 \wedge x_2 \leq t_1 \equiv t_2$

(2) $x_1 \wedge x_2' \leq t_1 \equiv t_2$

(3) $x_1' \wedge x_2 \leq t_1 \equiv t_2$

(4) $x_1' \wedge x_2' \leq t_1 \equiv t_2$

[In other words, for each of the four basic terms B in the variables x_1 and x_2, $B \leq t_1 \equiv t_2$ is provable].

Then $t_1 = t_2$ must be provable (from $B_1^*, B_2 - B_5$, because from (1) and (2) it follows by B_5(a) that $x_1 \leq t_1 \equiv t_2$. Then, from (3) and (4) it similarly follows that $x_1' \leq t_1 \equiv t_2$. Then from B_5(b) we get $t_1 = t_2$.

Next, let's consider basic terms in 3 variables x_1, x_2 and x_3. Suppose t_1 and t_2 are terms such that for each basic term B in the variables x_1, x_2 and x_3, the equation $B \leq t_1 \equiv t_2$ is provable. Thus the following eight things are provable.

(1) $x_1 \wedge x_2 \wedge x_3 \leq t_1 \equiv t_2$

(2) $x_1 \wedge x_2 \wedge x_3' \leq t_1 \equiv t_2$

(3) $x_1 \wedge x_2' \wedge x_3 \leq t_1 \equiv t_2$

(4) $x_1 \wedge x_2' \wedge x_3' \leq t_1 \equiv t_2$

(5) $x_1' \wedge x_2 \wedge x_3 \leq t_1 \equiv t_2$

(6) $x_1' \wedge x_2 \wedge x_3' \leq t_1 \equiv t_2$

(7) $x_1' \wedge x_2' \wedge x_3 \leq t_1 \equiv t_2$

(8) $x_1' \wedge x_2' \wedge x_3' \leq t_1 \equiv t_2$

From (1) and (2) we get: (a) $x_1 \wedge x_2 \leq t_1 \equiv t_2$

From (3) and (4) we get: (b) $x_1 \wedge x_2' \leq t_1 \equiv t_2$

From (5) and (6) we get: (c) $x_1' \wedge x_2 \leq t_1 \equiv t_2$

From (7) and (8) we get: (d) $x_1' \wedge x_2' \leq t_1 \equiv t_2$

Then from (a), (b), (c) and (d), we get $t_1 = t_2$, as we saw before.

The same sort of argument also applies to basic terms of 4 or more variables (if the argument works for n, it also works for $n+1$ in essentially the same manner as the above way we reduced the case of 3 variables to the case of 2 variables, and so by mathematical induction, the argument works for all n). We thus have:

B_{10}: For any positive number n and any terms t_1 and t_2, if each of the 2^n basic terms B, the equation $B \leq t_1 \equiv t_2$ is derivable from $B_1 - B_5$, then $t_1 = t_2$ is derivable from $B_1 - B_5$.

Now we can wind up the matter. Suppose $t_1 = t_2$ is a Boolean tautology. Let $x_1, \ldots, x_n$ be the variables that occur in $t_1 = t_2$. Then, for each of the 2^n basic terms B in the variables $x_1, \ldots, x_n$, we have $B \leq t_1 \equiv t_2$ derivable from $B_1 - B_5$

(by law B_9). Hence, by law $B_{10}, t_1 = t_2$ is derivable from $B_1 - B_5$. Thus, $B_1 - B_5$ is an adequate postulate system for Boolean algebras.

Let us illustrate with an example. Suppose we wish to show that the tautological equation $x \supset y = y' \supset x'$ is derivable from $B_1 - B_5$. [Remember, we have already derived laws B_6 and B_7, so we are free to use them]. Well, let t be the term $(x \supset y) \equiv (y' \supset x')$. We must show that $B \leq t$ for each basic term B in x and y. We can conveniently do this using the following table, which is essentially like a truth table.

0	1	2	3	4	5	6	7
	x	y	x'	y'	$x \supset y$	$y' \supset x'$	$(x \supset y) \equiv (y' \supset x')$
1 $x \wedge y$	+	+	-	-	+	+	+
2 $x \wedge y'$	+	-	-	+	-	-	+
3 $x' \wedge y$	-	+	+	-	+	+	+
4 $x' \wedge y'$	-	-	+	+	+	+	+

Explanation – A plus sign in any box indicates that $B \leq t$ when B is the basic term heading the row, and t is the term at the head of the column, whereas a minus sign indicates that $B \leq t'$. For example, the plus sign in the 3rd row, 5th column abbreviates the statement $x' \wedge y \leq x \supset y$, whereas the minus sign in the 2nd row, 6th column abbreviates the statement $x \wedge y' \leq (y' \supset x')'$. Each row is actually constructed according to the laws $B_1^*, B_2 - B_7$. For example, the seven signs of the 2nd row are simply an abbreviation of the following sequence of seven statements (whose justification, in terms of $B_1 - B_7$, is indicated alongside).

(1) $x \wedge y' \leq x$ (by B_1^*)
(2) $x \wedge y' \leq y'$ (by B_1^*)
(3) $x \wedge y' \leq x''$ (from (1), by B_4)
(4) $x \wedge y' \leq y'$ (repetition of (2))
(5) $x \wedge y' \leq (x \supset y)'$ (from (1) and (2) by B_6)
(6) $x \wedge y' \leq (y' \supset x')'$ (from (4) and (3) by B_6)
(7) $x \wedge y' \leq (x \supset y) \equiv (y' \supset x')$ (from (5) and (6) by B_7)

The resemblance of the above table to a truth table is more than casual: If we replace each + by T and each minus by F and ignore the 0 column, and regard x and y as propositional variables instead of Boolean variables, we literally get a truth table!

III – The Boolean Ring Approach

We have remarked that there exist many postulate systems for Boolean algebra. We now turn to another one which is particularly neat and is closely related to the algebraic approach to propositional logic that we considered in Part III of Chapter 17.

We now consider a non-empty set ∪, together with an operation that assigns to each element x and y an element denoted xy, and another operation that assigns to x and y an element denoted x+y, and two distinct elements of ∪ denoted 1 and 0. The collection of these items is called a *Boolean Ring* if for all elements x, y, and z of ∪, the following laws hold:

R_1: $xy = yx$

R_2: $x(yz) = (xy)z$

R_3: $x + y = y + x$

R_4: $x + (y + z) = (x + y) + z$

R_5: $x(y + z) = xy + xz$

R_6: $x + 0 = x$

R_7: $x1 = x$

R_8: $xx = x$

R_9: $x + x = 0$

[We remark that if we take ∪ to be the set of natural numbers and we take xy to be x times y and x+y to be x plus y, then $R_1 - R_7$ all hold but, of course, R_8 and R_9 don't. However, if we reinterpret "=" to mean "has the same parity as" (*i.e.* both even or both odd), then R_8 and R_9 hold as well].

Now, suppose we start out with a Boolean algebra. If we now take xy to be $x \wedge y$, and x+y to be $(x \equiv y)'$ or alternatively $(x \wedge y') \vee (x' \wedge y)$, which is the same thing, then $R_1 - R_9$ all hold, since they are then tautological (by virtue of the tautologies $S_1 - S_9$ ending Part III of Chapter 17). Thus, under this interpretation of xy and x+y, we get a Boolean ring from a Boolean algebra.

What interests us more now, is that if we start out with a Boolean ring and then define $x \wedge y$ to be xy; $x \vee y$ to be $xy + (x + y)$; and x' to be x+1, we then get a Boolean algebra! One way to see this is to show that the postulates $P_1 - P_9$ starting this chapter are then derivable (which is quite easy), and we have already shown that $P_1 - P_9$ is an adequate postulate system. [Alternatively, one can show the adequacy of this system *directly* (in much the same manner as what we did in Part III of Chapter 17), and then show that $R_1 - R_9$ are derivable from $P_1 - P_9$, thus getting a second proof of the adequacy of $P_1 - P_9$].

Suppose now, we start out with a set ∪ with operations $\wedge, \vee$ and $'$, together with an element 0 and define 1 to be $0'$; xy to be $x \wedge y$; $x+y$ to be $(x \wedge y') \vee (x' \wedge y)$; and take as postulates $R_1 - R_9$ together with the following two:

R_{10}: $x' = x + 1$

R_{11}: $x \vee y = xy + (x + y)$

We then have another adequate postulate system for Boolean algebra.

Exercise 1 – Starting with a Boolean ring and defining $x \wedge y$ to be xy and $x \vee y$ to be $xy + (x + y)$ and x' to be $x + 1$, derive the postulates $P_1 - P_9$.

Exercise 2 – Conversely, starting with operations $\wedge, \vee$ and $'$ and defining xy to be $x \wedge y$; $x + y$ to be $(x \wedge y') \vee (x' \wedge y)$, derive $R_1 - R_9$ from $P_1 - P_9$.

For the reader who is interested, two other postulate systems can be found in the books by Rosenbloom and Bell and Slomson, referred to at the end of the last chapter.

SOLUTIONS

1 – It will be helpful to first prove the following law which we will call P_o : $x \vee y = (x' \wedge y')'$. Well, $(x \vee y)' = x' \wedge y'$ (by P_4), hence $(x \vee y)'' = (x' \wedge y')'$, but $(x \vee y)'' = x \vee y$ (by P_5), hence $x \vee y = (x' \wedge y')'$.
    Now, we shall derive $P_1^\circ - P_9^\circ$

P_1°: $x \vee y = (x' \wedge y')'$ (by P_o)

$$= (y' \wedge x')' \text{ (since } x' \wedge y' = y' \wedge x' \text{ by } P_1)$$
$$= y \vee x \text{ (by } P_o)$$

P_2°: $x \vee (y \vee z) = (x' \wedge (y \vee z)')'$ (by P_o)

$$= (x' \wedge (y' \wedge z'))' \text{ (since } (y \vee z)' = y' \wedge z', \text{ by } P_4)$$
$$= ((x' \wedge y') \wedge z')' \text{ (by } P_2)$$
$$= ((x \vee y)' \wedge z')' \text{ (since } x' \wedge y' = (x \vee y)' \text{ by } P_4)$$
$$= (x \vee y) \vee z \text{ (by } P_4)$$

P_3°: $x \vee (y \wedge z) = (x' \wedge (y \vee z)')'$ (by P_o)

$$= (x' \wedge (y' \wedge z'))' \text{ (since } (y \vee z)' = y' \wedge z' \text{ by } P_4)$$
$$= ((x' \wedge y') \vee (x' \wedge z'))' \text{ (by } P_3)$$
$$= (x' \wedge y')' \wedge (x' \wedge z')' \text{ (by } P_4)$$
$$= (x \vee y) \wedge (x \vee z) \text{ (since } ((x' \wedge y')' = x \vee y \text{ and } (x' \wedge z')' = x \vee z, \text{ by } P_4)$$

P_4°: Since $x = x''$ and $y = y''$ (by P_5) then

$$(x \wedge y)' = (x'' \wedge y'')' = x' \vee y' \text{ (by } P_4). \text{ [To see this more clearly, let } z = x'$$
$$\text{and } w = y'. \text{ Then } (x'' \wedge y'')' = (z' \wedge w')' = z \vee w = x' \vee y'].$$

P_5°: This is the same as P_5. [P_5 is its own dual].

P_6°: $x \vee x' = (x' \wedge x'')'$ (by $P_\circ$)

$\qquad = (x' \wedge x)'$ (since $x'' = x$ by P_5)

$\qquad = (x \wedge x')'$ (P_1)

$\qquad = 0'$ (since $x \wedge x' = 0$ by P_6)

$\qquad = 1$ (by P_9)

To avoid duplication of labor, we will do P_9° before P_7°.

P_9° : $1' = 0''$ (since $1 = 0'$ by P_9) $= 0$) (by P_5).

P_7°: $x \vee 1 = (x' \wedge 1')'$ (by $P_\circ$)$= (x' \wedge 0)'$ (since $1' = 0$ by P_9°)

$\qquad = 0'$ (since $x' \wedge 0 = 0$ by P_7)$= 1$ (by P_9)

P_8°: $x \vee 0 = (x' \wedge 0')'$ ($P_\circ$) $= (x' \wedge 1)'$ (since $0' = 1$ by P_9)

$\qquad = (x')'$ (since $x' \wedge 1 = x'$ by P_8) $= x$) (by P_5)

$2 - Q_1$: If $x = y$, then of course $x' = y'$. Now suppose $x' = y'$, then $x'' = y''$, hence $x = y$ (since $x'' = x$ and $y'' = y$ by P_5).

Q_2: (a) $(x \wedge y) \vee (x \wedge y') = x \wedge (y \vee y')$ (P_3)

$\qquad = x \wedge 1$ (since $y \vee y' = 1$ by P_6°)

$\qquad = x$ (by P_8)

(b) This is the dual of (a)

Q_3: (a) We first show that (1) is equivalent to (2). Well, suppose $x \wedge y = x$. Then $(x \wedge y) \wedge y' = x \wedge (y \wedge y')$($P_2$) $= x \wedge 0$ (by P_6)$= 0$ (by P_7). Thus, $x \wedge y = x$ implies $x \wedge y' = 0$. Conversely, suppose $x \wedge y' = 0$. Since $(x \wedge y) \vee (x \wedge y') = x$ (by Q_2(a)) and $x \wedge y' = 0$ (by assumption), then $(x \wedge y) \vee 0 = x$. But $(x \wedge y) \vee 0 = x \wedge y$ (by P_8°), and so $x \wedge y = x$.

Thus, $x \wedge y = x$ is equivalent to $x \wedge y' = 0$

(b) From (a), it follows by the second duality principle (Proposition 2) that $x \vee y = x$ iff $x \vee y' = 1$. Hence by P_1°, $y \vee x = x$ iff $y' \vee x = 1$. Interchanging x with y, we have $x \vee y = y$ iff $x' \vee y = 1$.

(c) We now know that (1) is equivalent to (2) and that (3) is equivalent to (4). It remains to show that (2) is equivalent to (3), which will then make all four conditions equivalent. Well, $x \wedge y' = 0$ iff $(x \wedge y')' = 0'$ (by Q_1) iff $(x' \vee y'') = 1$ (by P_4° and P_9) iff $(x' \vee y) = 1$ (by P_5).

Q_4: $x \leq y$ iff $x \vee y = y$ (Q_3) iff $(x \vee y)' = y'$ (Q_1) iff $x' \wedge y' = y'$ (P_4) iff $y' \wedge x' = y'$ (P_1). Thus $x \leq y$ iff $y' \leq x'$.

Q_5: By Q_3, taking x for y, $x \wedge x = x$ iff $x \wedge x' = 0$. But $x \wedge x' = 0$ (P_6), hence $x \wedge x = x$–or in the alternative notation, $x \leq x$.

Q_6: (a) Using P_1 and P_2, $(x \wedge y) \wedge x = x \wedge (x \wedge y) = (x \wedge x) \wedge y = x \wedge y$ (since $x \wedge x = x$ by Q_5). Thus $(x \wedge y) \wedge x = x \wedge y$, which means that $x \wedge y \leq x$.
    (b) By (a), $y \wedge x \leq y$, hence $x \wedge y \leq y$ (by P_1).
    (c) By (a), $x' \wedge y' \leq x'$, hence by P_4, $(x \vee y)' \leq x'$, hence $x \leq x \vee y$ (by Q_4).
    (d) By (c), $y \leq (y \vee x)$, hence $y \leq x \vee y$ (by P_1°).

Q_7: (a) This says nothing more than that $x \leq x$, which is Q_5.
    (b) This follows from P_5 and (a).
    (c) This follows from P_4 and (a).

Q_8: Suppose $x \leq y$ and $y \leq z$. Thus $x \wedge y = x$ and $y \wedge z = y$. Since $y = y \wedge z$, then $x \wedge y = x \wedge (y \wedge z)$. Also $x \wedge (y \wedge z) = (x \wedge y) \wedge z$ (P_2). Thus $x \wedge y = (x \wedge y) \wedge z$, but also $x = x \wedge y$, hence $x = x \wedge z$ so $x \wedge z = x$, which means $x \leq z$.

Q_9: (a) Suppose $x \leq y$ and $x \leq z$. Thus $x \wedge y = x$ and $x \wedge z = x$. Since $x = x \wedge y$, then we can replace x by $x \wedge y$ in the equation $x \wedge z = x$, thus obtaining $(x \wedge y) \wedge z = x$. Hence $x \wedge (y \wedge z) = x$ (P_2) and so $x \leq y \wedge z$.
    (b) Suppose $x \leq z$ and $y \leq z$. Thus $x \wedge z = x$ and $y \wedge z = y$. Then by Q_3, $x \vee z = z$ and $y \vee z = z$. Since $z = y \vee z$, then $x \vee z = x \vee (y \vee z) = (x \vee y) \vee z$ (P_2°). Hence $(x \vee y) \vee z = z$, so $x \vee y \leq z$ (by Q_3).

Q_{10}: Suppose $x \wedge y \leq z$ and $x \wedge y' \leq z$. Then $(x \wedge y) \vee (x \wedge y') \leq z$ (by Q_9(b)). Then $x \wedge (y \vee y') \leq z$ (by P_3). Hence $x \leq z$ (since $y \vee y' = 1$, by P_6°) and $x \wedge 1 = x$, by P_8).

Q_{11}: Suppose $x \leq y$ and $y \leq x$. Then $x \wedge y = x$ and $y \wedge x = y$. Since $y \wedge x = y$, then $x \wedge y = y$ (by P_1). Thus $x \wedge y = x$ and $x \wedge y = y$, so $x = y$.

Q_{12}: (a) $x \leq 1$ means $x \wedge 1 = x$, which is true by P_8.
    (b) Suppose $1 \leq x$. Since also $x \leq 1$ (by (a), then $x = 1$ (by Q_{11}).
    (c) Since $x \wedge 0 = 0$ (P_7) and $x \wedge 0 = 0 \wedge x$ (P_1), then $0 \wedge x = 0$, and thus $0 \leq x$.
    (d) Suppose $x \leq 0$. Since also $0 \leq x$ (by (c)), then $x = 0$ (by Q_{11}).

Q_{13}: Suppose $x \leq z$ and $x' \leq z$, then $x \vee x' \leq z$ (by Q_9(b)), hence $1 \leq z$ (by P_6°), and so $z = 1$ (by Q_{12}(b)).

Q_{14}: To begin with, $(x \equiv x) = 1$, because $x \equiv x = (x' \vee x) \wedge (x' \vee x) = (x \vee x') \wedge (x \vee x')$ (by P_1°)= $1 \wedge 1$ (by P_6°)= 1 (by P_8). Therefore, if $x = y$, then $(x \equiv y) = 1$. Conversely, suppose $(x \equiv y) = 1$. Thus $(x \supset y) \wedge (y \supset x) = 1$. Now, $(x \supset y) \wedge (y \supset x) \leq x \supset y$, and $(x \supset y) \wedge (y \supset x) \leq y \supset x$ (by Q_6(a) and (b)). Thus $1 \leq x \supset y$ and $1 \leq y \supset x$. Hence $x \supset y = 1$ and $y \supset x = 1$ (by Q_{12}(b)), and therefore $x \leq y$ and $y \leq x$ (by Q_3), and so by Q_{11}, $x = y$.

Q_{15}: Suppose $x \leq y \equiv z$ and $x' \leq y \equiv z$, then $y \equiv z = 1$ (by Q_3). Hence $y = z$ (by Q_{14}).

3 – B_1 (a) This is L_1.

(b) Suppose $x \leq y$. Now $x \wedge z \leq x$ (L_2). Also, $x \leq y$, so $x \wedge z \leq y$ (by L_8). Similarly $z \wedge x \leq x$ (L_3), so $z \wedge x \leq y$ (by L_8).

B_2: (a) This is L_9.

(b) By $L_2, y \wedge z \leq y$, hence by $L_{10}, y' \leq (y \wedge z)'$. Therefore if $x \leq y'$, then $x \leq (y \wedge z)'$. Similarly, if $x \leq z'$ then $x \leq (y \wedge z)'$, using L_3 instead of L_2.

B_3: (a) Suppose $x \leq y$. Since also $y \leq y \vee z$ (by L_4) then by $L_8, x \leq y \vee z$. Similarly, if $x \leq z$, then $x \leq y \vee z$ (using L_5 instead of L_4).

(b) Suppose $x \leq y'$ and $x \leq z'$. Then $x \leq (y' \wedge z')$ (by L_9). Also $(y' \wedge z') \leq (y \vee z)'$ (L_6). Therefore $x \leq (y \vee z)'$ (by L_8).

B_4: Suppose $x \leq y$. Since also $y \leq y''$ (by L_7) then by L_8, it follows that $x \leq y''$.

B_5: (a) This is L_{11}.

(b) This is L_{12}.

4 – Let B be a basic term $x_1^* \wedge \ldots \wedge x_n^*$ in the variables $x_1 \ldots x_n$.

Step 1 – For any i from 1 to n, x_i^* is either x_i or x_i'. If the former, then $B \leq x_i$, by B_1^*. If the latter, then $B \leq x_i'$, again by B_1^*, and so either $B \leq x_i$ or $B \leq x_i'$, and so B governs x_i. Thus B governs each of the variables $x_1, \ldots, x_n$.

Step 2 – Suppose that B governs terms t_1 and t_2. We are to show that B governs $t_1 \wedge t_2$ and $t_1 \vee t_2$, and also that if B governs a term t, then B governs t'. Well, suppose B governs both t_1 and t_2, then there are 4 possibilities:

(1) $B \leq t_1$ and $B \leq t_2$

(2) $B \leq t_1$ and $B \leq t_2'$

(3) $B \leq t_1'$ and $B \leq t_2$

(4) $B \leq t_1'$ and $B \leq t_2'$

If (1) holds, then $B \leq t_1 \wedge t_2$ by B_2(a). If either (2), (3) or (4) holds, then either $B \leq t_1'$ or $B \leq t_2'$, hence by B_2(b), we have $B \leq (t_1 \wedge t_2)'$. Thus either $B \leq t_1 \wedge t_2$, or $B \leq (t_1 \wedge t_2)'$, and so B governs $t_1 \wedge t_2$.

The proof that B governs $t_1 \vee t_2$ is quite similar using B_3 instead of B_2.

Now, suppose that B governs t. Then either $B \leq t'$ or $B \leq t$. If $B \leq t$, then $B \leq t''$ (by B_4). Thus either $B \leq t'$ or $B \leq t''$, and so B governs t'.

By mathematical induction, it follows that for *every* term t in the variables $x_1, \ldots, x_n$, B governs t. [As an example of an application, suppose B is a basic term in the variables x_1, x_2 and x_3. Suppose we wish to show that B governs the term $(x_1 \wedge x_2')' \vee x_3$. Well, B does govern x_1, x_2 and x_3 (as we have seen). Then B successively governs x_2', then $x_1 \wedge x_2'$, then $(x_1 \wedge x_2')'$ and finally $(x_1 \wedge x_2')' \vee x_3$].

5 – (a) Consider a basic term $x_1^* \wedge \ldots \wedge x_n^*$. For each number i from 1 to n, either x_i^* is x_i, or x_i^* is x_i'. If x_i^* is x_i and we take the value 1 for x_i, then $x_i^* = 1$. If x_i^* is x_i', then again x_i^* can be made to have the value of 1 by giving x_i the value of 0. And so by substituting 0's and 1's for $x_1, \ldots, x_n$, each x_i^* can be made to have the value of 1, hence $x_1^* \wedge \ldots \wedge x_n^*$ will then have the value $1 \wedge \ldots \wedge 1$, which is 1. Thus, $x_1^* \wedge \ldots \wedge x_n^*$ is not 0 for *all* values of $x_1, \ldots, x_n$ (for some values it is 1).

(b) Suppose $B \leq t$ and $B \leq t'$ are both valid. Then so is $B \leq (t \wedge t')$ (by B_1). But $t \wedge t'$ is identically 0 (since $t \wedge t' = 0$ is tautological), hence $B \leq 0$, for all values of the variables in B, hence $B = 0$ for all values of the variables in B, and thus B is identically 0. Then B cannot be basic (by (a)). Thus if B is basic, then $B \leq t$ and $B \leq t'$ cannot both be valid.

(c) Suppose that B is basic and that t is a term whose variables are included in the variables of B. Then B governs t (by B_8), and so either $B \leq t$ or $B \leq t'$ is provable from $B_1^*, B_2 - B_5$. Suppose also that t is identically 1. Then t' is identically 0, hence if $B \leq t'$ were provable, it would be valid, hence $B \leq 0$ would be valid, which would mean that B was identically 0, contrary to (a). Thus, $B \leq t'$ is not provable, so it must be $B \leq t$.

6 – Suppose $t_1 = t_2$ is tautological. Then it is valid, hence $t_1 \equiv t_2$ is identically 1. Then by (c) of Problem 5, $B \leq t_1 \equiv t_2$ is provable for every basic term B in the variables of $t_1 \equiv t_2$.

Chapter 22

George Boole and Mathematical Logic

I – George Boole and His Work

Boolean algebras play a key role today, not only in the foundations of mathematics, but also in computer science and the design of electrical networks. The father of Boolean algebra was the nineteenth century mathematician, George Boole (1815-1864), whose purpose was to reformulate logic in mathematical terms and who was one of the first to apply algebraic symbolism to logic. Indeed, modern mathematical logic may be said to have begun with Boole. His most famous work is: "An Investigation of Laws of Thought, on which are founded the Mathematical Theories of Logic and Probabilities" (London, 1854, reprinted by Dover Publications). Boole described his purpose in writing this book in the following opening words:

> The design of the following treatise is to investigate the fundamental laws of the mind by which reasoning is performed; to give expression to them in the symbolic language of a Calculus, and upon this foundation to establish the science of Logic and construct its method; to make that method itself the basis of a general method for the application of the mathematical doctrine of Probabilities; and, finally, to collect from the various elements of truth brought to view in the course of these inquiries some probable intimations concerning the nature and constitution of the human mind.

Boole had the remarkable distinction of being self-educated, but who, nevertheless, was one of the leading mathematicians of his generation. His father, a shoemaker, could afford only a simple schooling for him; after that he studied from books. He easily learned Latin and Greek, and then went on to higher mathematics. In addition to his pioneering work in logic, he also made contributions to the fields known as *differential equations* and *Finite Differences*. From the age of 16 to the age of 34, Boole was a schoolmaster, much of the time at a school he ran himself, but during all this time, he steadily pursued his mathematical studies. He gradually built up a considerable reputation by a series of mathematical papers and was awarded a medal in 1844 by the Royal Society of London for his fundamental paper titled: "On a General Method in Analysis". In 1849 he was appointed to the chair of mathematics at Queen's College at Cork, where he remained until his death in 1864. His later work in logic was done during these years in Ireland.

His *Laws of Thought* is a curious mixture of precise mathematical and symbolic reasoning with philosophical considerations. He attempts to put purely philosophical arguments into symbolic form–especially in his chapter on the philosophers Clarke and Spinoza. Towards the beginning of this chapter, he says the following:

> In the pursuit of these objects it will not devolve upon me to inquire, except incidentally, how far the metaphysical principles laid down in these celebrated productions are worthy of confidence, but only to ascertain what conclusions may justly be drawn from the given premises.

In other words, Boole's purpose in this chapter is not to decide whether the philosopher's premises (and hence also conclusions) are *true*, but only whether the conclusions really follow logically from the premises. And so here he makes a clear distinction between what are now called *valid* arguments and *sound* arguments. A *valid* argument is one in which the conclusion logically follows from the premises, regardless of whether the premises themselves are true or not; whereas a *sound* argument is one which is not only valid, but is such that the premises (and also the conclusion) are true. For example, the following argument is sound:

> All men are mortal.
> *Socrates is a man.*
> ∴ Socrates is mortal.
>
> [The symbol ∴ abbreviates "therefore"].

The following argument, though clearly not sound, is nevertheless valid.

> All men have two heads.
> *Socrates is a man.*
> ∴ Socrates has two heads.

Clearly, the first premise is false (as far as I know!) and so the argument is not sound, but it definitely is valid (since if it were really true that all men have two heads and that Socrates is a man, then Socrates would indeed have to have two heads).

There are some arguments, incidentally, that appear to be obviously invalid, but which in fact are valid. One of my favorites is the following:

> Everyone loves my baby.
> *My baby loves only me.*
> ∴ I am my own baby.

Printed in the United States
By Bookmasters